Making Space:
Strategic Leadership for a Complex World

Dr. Wanda M. Austin

The Aerospace Press
El Segundo, California

The Aerospace Press
2310 E. El Segundo Boulevard
El Segundo, California 90245-4691

ISBN-10: 1-534878-18-1
ISBN-13: 978-1-534878-18-1

DEDICATION

To the future leaders…

Aaliyah, Elianna, Mya, and Malachi.

ACKNOWLEDGMENTS

Wade Austin, Jr. has been the wind beneath my wings for forty-two years. Without his support and personal sacrifice, this book, and many other projects, would not have been possible. Wade, thanks so very much for your unconditional love.

My mother and father were my first role models. I'd like to thank them for many things, but most of all, for their courage.

There are many teachers who have influenced me over the years, but two of them have had a truly life-altering impact on me. One is Gary Cohen, who, when I was in seventh grade, told me that I was good at math and then worked with me until I believed it. The other is George Rosenstein, who took the time to scold me for skipping his 8 a.m. calculus class, then hired me to babysit his daughter so that I could earn extra money, regularly invited me to his home for dessert and pep talks, and strongly advised me to attend graduate school—which, at that time, wasn't something I had previously considered.

Special thanks to Stephen Presley, my collaborator and indispensable writing partner. Our leadership discussions began in 2011, while Steve was finishing his doctoral dissertation, *How Leaders Engage in Complexity Leadership: Do Action-Logics Make a Difference?* His perspectives on organizational effectiveness were extremely valuable in the development of the structure of this book. Our conversations, and the hours of recordings he carefully sifted through, enabled us to develop key themes that would serve as the foundation of the book. Thanks also to Karen Presley for contributing editorial and moral support.

There isn't enough room in this book to individually thank and express my gratitude to all of my fellow employees, industry colleagues, and customers who contributed to my personal and professional development and taught me valuable lessons over the years. To all of you, I extend my deepest thanks.

Special thanks to Jane Harman, Barbara Barrett, and Mike J. Daugherty for allowing me to watch and learn from their example; Dr. David Gorney for bringing the inexhaustible bench strength; and Larcine Gantner, a friend and amazing executive assistant throughout my tenure as CEO and President.

Thanks to Matt Kivel (writer, editor, relentless taskmaster), Lindsay Chaney (editor, organizer), Rick Humphrey (cover artist), Shane Glaseman (editor), and the production team (Denise Betts, Barney Sasaki, and Mary Villanueva), who were all ably led by Sabrina Steele. Their enthusiasm, dedication, and never-ending energy kept the project on schedule and produced excellent results.

Contents

I know a thing or two about boys' clubs. At Harvard Law School, I was one of 30 women in a class of more than 500. Throughout my career—on Capitol Hill, at the White House, and at the Department of Defense—I was the rare woman at the table. Back then, few shared my view that national security issues are women's issues, and vice versa. I was proud, though, to meet and work with talented women who agreed. Many of them have gone on to exceptional careers in public service or business leadership; they remain dear friends and role models.

One impressive dear friend is Wanda Austin, a *bona fide* rocket scientist and a true rock star. During my nine terms in Congress, I represented the aerospace heart of California; my talented constituents made the most of the nation's sensitive intelligence satellites. As head of The Aerospace Corporation—the federally funded research and development center in my district—Wanda stood out.

Wanda has confidence, competence, grit; her work ethic is incomparable; and she can take a punch. And as this book makes clear, she's committed to helping those still climbing the ranks.

Every audience—men and women, young and not so much—has something to learn from Wanda Austin.

Different parts of her journey will speak to different readers, but I'll say this: her experience resonates deeply with the lessons I've learned over a long career in national security. I've learned that leadership works inside-out, starting with your own head and heart. I've seen that leadership takes work, a discipline Wanda has in spades. I've seen that failure can be a friend, that leaders never give up, and that leadership—especially for women—can be lonely.

Wanda knows that leaders live real lives. We're all more than the sum of our job titles; we all gain strength from spouses and colleagues, families and mentors.

Making Space: Strategic Leadership for a Complex World isn't a handbook for super-heroes or one-woman-armies. It's a blueprint for concrete, real-world leadership.

— *Jane Harman*
  President and CEO
  Woodrow Wilson International Center for Scholars

INTRODUCTION

It was the middle of the 1960s and I was living in New York City—not far from Yankee Stadium. I was 11 years old. My family was tight-knit, supportive… and unified in its belief that academic achievement was the key to success. My parents always encouraged me to find a path in life. A path to higher education. A path to a great career. A path out of our neighborhood. Although I heeded my parents' advice and focused on my schoolwork, I was always unsure of the path I needed to take. I studied hard and flourished during my early years, but always, something was missing. I couldn't figure out what all of this time and effort in school was leading me toward. There were few examples of academic success in my neighborhood and even fewer sources of inspiration for career opportunities for young, African American women like myself. The examples I found in the media were very unsatisfactory, making it difficult to answer the question, "What do you want to be when you grow up?"

Some people know early on what they want to be when they grow up. These people, for whatever reason, have found their calling at a young age. They might have a family business that they'd like to join. They might recognize their individual talents and devote themselves to specific careers that perfectly match their abilities. They might have a family member whose career and life they would like to emulate. Whatever the motivation, these people are uniquely focused and self-assured. They know what they want and they devote themselves to achieving their chosen pursuits.

Growing up, I was not one of those people. I didn't know where my journey in life would take me. I was always searching among the options. I was always looking for a passion that I could eventually turn into a profession, but I didn't know where to begin. Until, that is, I entered junior high school. Walking into Paul Hoffman Junior High School in The Bronx, my calling suddenly became a little bit clearer and my path in life, unbeknownst to me, started to take shape. It started with a math teacher.

It was the spring of 1967 and I was 12 years old, sitting nervously in the front of Mr. Cohen's seventh-grade math class. Mr. Cohen walked down between the rows of desks, passing out a very challenging math problem to all of the students. Over the next few minutes, I worked diligently on the problem, solving it with relative ease. When I was done, I looked around the room to see many of my fellow students still working away, or talking amongst themselves, having given up. The next day, Mr. Cohen returned the graded problem. As he did, he made a very surprising comment that the whole class heard. He singled me out and said "Wanda, you're good at this and don't ever let anyone tell you that you're not." His words that day, and the encouragement and mentoring he provided in the ensuing years, would have a profound impact on my life. He gave me the confirmation I needed, a direction and a goal to work toward. He enabled me to believe in myself and helped me build a foundation for the engineering career that I would embark upon during my college years and beyond.

That wasn't the last time a teacher helped me to focus on my particular strengths. Four years later, I was spending some free time in the coffee shop at Franklin and Marshall College, laughing and relaxing with friends. School, and math in particular, had become my sanctuary, my second home. In such an atmosphere, a person can tend to relax a bit more than might be advisable. I learned this when Professor Rosenstein—my calculus professor—suddenly appeared and tapped me on the shoulder. He looked me in the eyes and said, "You were not in my class today. Don't ever let that happen again." I was startled. I didn't think he had even noticed my absence, and I certainly hadn't expected to be criticized publicly for not showing up. His words stung me, but I realized he was right. I had an amazing opportunity to get a college education and I couldn't afford to squander it. My talent could only get me so far. I needed to put the work in. Needless to say, I never missed another class; not his or anyone else's, and I went on to graduate from Franklin and Marshall with a bachelor's degree in mathematics.

Professor Rosenstein's wake-up call kept me focused for a long time. Twelve years later, I found myself in Southern California, working as a project engineer at The Aerospace Corporation. My job was wonderful. I enjoyed interacting with my colleagues and customers and I was thrilled to support Aerospace's national security mission. There were new engineering challenges and learning opportunities every day, and I absolutely loved my work. After a time, though, that same old question began to gnaw at me again—what do you want to be when you grow up? I couldn't believe that I was still asking myself that question. After all, I already *was* grown up, wasn't I? How could I still be looking for my purpose? I felt satisfied with what I had accomplished professionally, yet I knew, undoubtedly, that I hadn't reached my potential from a career standpoint. Eberhardt Rechtin, president of The Aerospace Corporation, recognized that I could do more.

During a cocktail party, he suggested I go back to school and get a Ph.D. in engineering. He talked with me at length about exciting career opportunities that would be available to me if I had my doctorate. I listened intently and, while I took the next few weeks to carefully consider his advice, I knew what I needed to do. Before long I was enrolled at the University of Southern California, and several years later I graduated with a Ph.D. in industrial and systems engineering. Earning my degree, especially while balancing work and family, was a very satisfying accomplishment, but I knew that I could do even more. I was an adult with a career, a great husband, a young family, and a passion for engineering, and still I didn't feel like I had it all figured out. Something was missing. That same question came back once again: *What do you want to be when you grow up?*

Seven years later, I became a principal director at Aerospace. This was an advanced leadership position that required me to manage budgets, projects, people, and personalities. I no longer had the luxury of focusing solely upon engineering problems—I had to make the big decisions that enabled the engineering work to get done. My organization consisted of approximately 100 engineers and scientists, all supporting both a wide range of engineering products and a diversity of customers. While this new role was unfamiliar to me, it felt right. I felt it bringing me closer to what I had been looking for all of those years ago in The Bronx. Then, one day, it all came together for me.

That day, a small team presented me with a project that had hit several roadblocks: the customer expectations were unclear, the team lacked access to specific required expertise, and they needed funding to conduct a critical test in the lab. We all met in my office and, after listening to their project status report, I identified three things that I could do to help them eliminate those barriers. I called the customer directly to clarify the expectations for the project and to establish a realistic timeline for results. I contacted the senior vice president of my organization to petition for discretionary dollars to be used for the lab tests that were needed. Finally, I contacted a colleague from another division to get an individual with the right expertise to work on the project for a short time. Once all this was accomplished, I called the team together to tell them the good news—their project was a "go." They were visibly elated as they left my office, excited that they had an executable path forward for getting their important work done.

My first response to their reaction was an emotional one, a sense of loss. I realized that, once I had eliminated their obstacles, they no longer needed my help and I would not be a part of the technical engineering effort to develop the final product. After a few minutes, though, I realized that what I had done was very useful for them, and incredibly rewarding to me—I had provided solid leadership for the team. My role was to enable them to be successful and I did just that. Of course, there is an undeniable satisfaction that comes with being an individual contributor, but for me, it was even more satisfying to know that I was able to empower others to do great things—well beyond what I could have accomplished on my own. The leadership I had shown filled me with a sense of newfound purpose. It felt like I had finally found my calling; through the help of my parents, my teachers, and my colleagues, I finally knew what I wanted to be when I "grew up," and I was already doing it. I wanted to lead.

In the ensuing years I would hone my craft and continue to perfect my methods of providing intentional and strategic leadership for my colleagues and customers. Time and time again, I found that although leadership is a constantly-evolving process and the work is always changing, there are certain fundamental principles that are absolutely essential to success. The problems change, but the values and strategic concepts that inform a sound leadership philosophy stay the same. Strategic leadership is not a random act. It requires preparation, continuous learning, humility, and compassion. Preparation enables you to be confident in your work. Continuous learning ensures that you recognize when things have changed, and allows you to adapt to new challenges. Humility is important because leadership is challenging and everyone—and I mean *everyone*—will make mistakes. Great leaders learn from their failures and are willing to share what they have learned. Compassion allows you to earn respect, loyalty, and mutual understanding from all of those you serve.

I learned—and continue to learn—all of these skills from my teachers, my colleagues, and my family. They are all important and I am forever indebted to all those who helped me along the way. Of all these lessons, though, I learned the greatest one from my father. My father, a barber, was an excellent leader and he taught me how to lead with compassion. He was the head of our family. Every Sunday when I was growing up, my whole family would eat dinner together. My father would always sit at the table on those nights with an empty plate in front of him until everyone else had finished their meals. Then, and only then, would he take food for himself. I once asked him why he did this, and he answered, "It's my job to feed you, so when you've had your fill, then I will eat." It was then that I realized his delayed meal was actually a symbolic gesture of deference to our family—the family he loved, the family he worked for, the family he served. He taught me that being a leader doesn't entitle you to anything, except to serve the people you lead. It's a lesson I have never forgotten and a lesson that guides my leadership approach to this day.

I've written this book to share some of the leadership lessons that I've learned during my incredible journey as a leader and CEO in the aerospace industry. No leader will have exactly the same experiences or face the same challenges that I have. This book is less about my personal story and more about what I've learned and what I can pass on. The experiences and perspectives on leadership that I have compiled here are designed to serve as a thought-provoking resource to guide anyone who is interested in improving or supplementing their individual approach to intentional and strategic leadership. It's a guide that contains all of the concepts that I have found to be absolutely fundamental to great leadership. The bottom line is that leadership is not accidental. You have to make the time and space with your team to be strategic in this complex and global world that we live in today.

It should be said that this book and the lessons it contains are not only for CEOs or senior leaders; they apply to, and can be applied by, anyone in a leadership role, and even those who are aspiring to one. No matter what "level" you're at, the leadership lessons contained herein can be useful.

Who This Book Is For

Sitting CEOs—CEOs are busy. It's the nature of the job. Many aren't making time for developmental reading, but they should be. Leadership education is a constant process. This book explores a number of challenges specific to the role.

Aspiring CEOs—You might not be there yet, but... someday. The CEO role presents myriad challenges that are difficult to prepare for and are often surprising. This book gives you a glimpse of what you might expect once you are on the job.

Board members—The job of a board is governance; to advise and guide the leaders of a company, particularly the CEO. Sometimes it can be very challenging for board members to relate to or understand the unique challenges that a CEO is facing. This book presents valuable information about the little-known, behind-the-scenes aspects of being a CEO, so that board members may provide more effective counsel to them. The book can also help guide board members through one of their most significant responsibilities—hiring a CEO.

Anyone who wants to be a better leader—We are all leaders in our own sphere of influence. It is important to be intentional in our personal and professional development. We all have customers and we all have bosses. It is a worthwhile goal to be able to delight both. We want to be the best leaders we can be. This book, though written from the perspective of a CEO, contains numerous lessons that can benefit leaders in any role.

Before you jump in and start benefitting from the lessons I've gathered together for you, I'd like to point out that we have a wider responsibility to ourselves as leaders and the workforce we lead—and that's the responsibility we have to leave things better than we found them. It's essential that we not only prepare ourselves and our teams to achieve success today, but we must also consider the workforce of the future. There are two important workforce issues that must be addressed by leaders that I am very passionate about: inclusion and STEM education.

Inclusion is an incredibly complex issue that requires a great deal of thoughtfulness and self-examination. Experience has proven to me that our society is much stronger, more capable, and innovative when we embrace diversity and inclusion. A strong leader will intentionally create systems and environments where people are engaged and diverse perspectives are sought out and acted upon to achieve mission success. Diversity and inclusion are leadership issues, but they are both broad and tremendously nuanced. Today's leaders must examine them seriously and prepare their organizations and workforce to understand and embrace the advantages of diversity and inclusion.

STEM education—the pursuit of careers in science, technology, engineering, and math—is in crisis in the United States. Fewer students than ever before in this country's history are competent in the subjects that will enable them to define and shape the future. Without strong, STEM-trained leaders, the U.S. will not be able to continue to compete on the world stage. It is a serious problem that is very near and dear to my heart, and it's essential that the leaders of today do all that they can to encourage our young people to prepare themselves for the needs of the future—both their own, and those of our society.

CHAPTER ONE

Leading the Organization

WHERE ARE YOU GOING, AND HOW ARE YOU GOING TO GET THERE?

From afar, leadership looks easy to many; it is most assuredly not. This opinion perhaps arises because we all have leadership roles in our daily lives, whether they result from our family relationships, community obligations, or interactions with colleagues at the office. We consider these leadership roles to be a normal part of our lives and thus nothing out of the ordinary. They are still not easy, but the challenges of these roles are simply a part of who we are.

The truth is that leadership opportunities are everywhere and are not exclusive to those with actual leadership titles. In fact, having a leadership title such as CEO, president, general manager, or director simply means that someone is holding you accountable for leading an organization. What all leadership roles have in common, however, is that if no one is following you, by definition, you are not leading.

One of the first leadership lessons that I learned in my career is that you can achieve very little by yourself. To achieve leadership success, you must be able to organize and manage a group of people to achieve a common goal. That's how progress is made.

The foundation for leading an organization is built by focusing on values, vision, strategy, and execution. This focus will enable you to respond to change, which is an inevitable part of life and business. The sources of change can be internal or external, anticipated or totally unexpected. If the values, vision, and strategy are understood and embraced by the entire organization, every decision at every level will be guided by them and will be aligned with the organization's goals.

The focus on execution creates measurable results and enables progress toward achieving these goals.

To be a successful leader, you must develop skills to clearly define the value of your work (why do it?) and the vision (where you're headed), in addition to being able to specify the strategic (long-term) and tactical (immediate-term) details of how to get there. These skills are developed throughout your life and a good leader is always looking to improve them.

Values guide you and inform every decision you make. They represent your most essential principles. Your vision is an aspirational goal, or set of goals, that can inspire and guide your organization over the course of many years; it should be adjusted at regular intervals to reflect changes in industry, the economy, and society as a whole. Your strategy positions your organization in the real world and charts its course in a way that emphasizes strengths and limits weaknesses. It is this preparation and planning that allows you to see your vision through.

Values, vision, and strategy combine to enhance and bring clarity to a CEO's leadership perspective. The CEO needs to have the broadest, and, simultaneously, the most detailed understanding of her industry and her company. Here's a quick illustrative example.

Imagine we live in a strange and fantastical world where every leader in every walk of life owns a very special camera, a "leadership camera." Each camera gives its owner a unique, specialized view of the work they must accomplish. Each camera body is the same, but each job requires a different lens in order to get the work done. For example, a space systems project leader working on the construction of a small portion of a satellite would need a tightly focused lens, because his job requires an incredible amount of attention to detail. However, a CEO's lens would need to be able to handle wide angles, microscopic zoom-ins, and telescopically broad zoom-outs. The greater flexibility of the CEO's lens reflects her need to be an expert in many areas, along with her need to be able to see things from many different perspectives. That's the nature of her job. It's a perfect, and sometimes maddening, hybrid. In the real world, sadly, there are no such thing as leadership cameras, but if there were, a CEO's would most certainly be some hybrid of the Hubble Space Telescope and a compound microscope.

THE POWER OF VALUES

Values define how a company operates on an innate level—shaping daily behavior and influencing decisions. Ideally, anyone in an organization can use its core values as a guide, particularly in the face of uncertainty or challenging circumstances. I say ideally, because it's not always easy for members of an organization, particularly a large one, to internalize values that can sometimes feel remote and broad. In a corporate setting, values need to be discussed and connections need to be consistently made between corporate values and corporate action. Without solid communication about values and their relevance in the everyday working world, they will feel irrelevant and/or inconsequential to everyone in the organization.

In the 55-year history of The Aerospace Corporation, the corporate values have remained the same, though their application, and the interpretation of them, has changed to reflect new contexts over time. The corporation's values are:

- Dedication to mission success
- Technical excellence
- Commitment to our people
- Objectivity
- Integrity

I'll admit, when I took over as CEO in 2008, those values were implied and not necessarily integrated into daily business and behavior. It had been a long time since Aerospace leadership had communicated about these terms and connected them to the work the company was actually doing. They existed in the corporate fabric, but I didn't know how much they would impact my own tenure as CEO. How naive I was! During the first few months on the job, I was amazed at how many times I found myself leaning on those half-century-old values for guidance and boundaries. My corporate values were inherited from predecessors who had left the company long before I took over as CEO, yet I was able to find strength and meaning in their time-tested tenets, over and over again.

Although the corporation's values were developed and tailored specifically for its space technology work, as you'll see, they're equally relevant to virtually any business or organization. Let's briefly examine the Aerospace corporate values.

Dedication to Mission Success

I'm sure that you've seen a rocket or shuttle launch on television, in the movies, or in person at some point in your life. At the very least, you've heard the familiar countdown: "three, two, one, liftoff!" Well, "dedication to mission success," Aerospace's first corporate value, means that when the countdown reaches "one" and mission control pushes the button for launch, the rocket actually works. It may sound simple enough, but there's a reason people commonly use the term "rocket science" when representing the antithesis of anything simple or easy—it takes a lot of brain power, work hours, and precision to ensure that a rocket gets off the ground and out of Earth's atmosphere.

At Aerospace, a mission failure is not a simple matter of figuring out what went wrong and fixing it. A launch failure means the rocket blew up—or was commanded to destruct—and in so doing, destroyed its payload, which was almost certainly a valuable satellite or other significant space asset. That failure can mean millions or billions of dollars lost and, more importantly, the loss of the capability the payload would have provided. I should also point out that this value, dedication to mission success, doesn't just apply to the rocket. Even a successful launch can result in a mission failure, if the satellite malfunctions in a significant way. Remember, these satellites are in space. We can't easily send a repairman out there to fix them—at least not yet. So it's Aerospace's objective to perform all of the critical mission assurance work before a rocket launch, on both the rocket and the payload it's lifting, to mitigate problems and minimize risk for any given project.

Mission assurance work is the cornerstone service that Aerospace provides and it presents an opportunity for the company to showcase its dedication to mission success on numerous occasions throughout each year. With every launch comes a new opportunity to problem-solve and make the essential recommendations Aerospace's customers need to get their spacecraft into space.

The idea of being dedicated to mission success is not just an Aerospace concept. Far from it, in fact. All companies have work that is absolutely essential to the success of their organization and it is that work that needs to be prioritized above all else. That means supporting your customers from the very beginning to the very end of every assignment. That means taking the time to check and re-check your data in order to produce a result that is accurate and reliable. "Dedication to mission success" means that whatever your objective is, in any industry, for any business, you always work hard to make sure that the job is done right.

Technical Excellence

Technical excellence is Aerospace's second value and it's one that can be interpreted many different ways. To me, it means that it's Aerospace's duty to stay at the vanguard of space technology. That means substantial investment in research and development, as well as in competitive hiring practices that allow the company to recruit and retain the best and the brightest minds in engineering and innovation. In addition, since many Aerospace customers are in the defense sector, the company's technical excellence must also be exhibited in its commitment to addressing security threats at every level, and to maintaining a firm grasp on the rapidly evolving nature of space and cybersecurity concerns. A commitment to technical excellence, regardless of industry or expertise, means that you are driven to operate at the highest level and remain at the cutting edge of your field.

Commitment to Our People

Aerospace's third value is commitment to our people. This doesn't just mean providing employees with a safe workplace, a good salary, and competitive benefits, though those things are important and are a part of the equation. Commitment to our people means that you create and provide an environment in which employees can succeed, where their ideas will be heard and employed, and where they have state-of-the-art tools to accomplish their mission. It means they feel confident that their efforts will be recognized and rewarded based on merit, and that their employer will stand with them when the going gets tough.

As the leader of an organization, it's very important that, when you're making decisions, you reflect on how those choices will impact your staff. You always need to keep your employees in mind, every step of the way. It is unlikely that you will ever make a decision that pleases everyone, but when people feel you are as dedicated to their success as they are, they will be that much more motivated to help you achieve your goals and, in turn, fulfill your corporate mission.

Objectivity and Integrity

Aerospace's fourth and fifth values, objectivity and integrity, work hand-in-hand—one cannot function without the other. Integrity and objectivity are essential qualities in every successful company because work that has a reputation of being trusted and reliable will flourish, and the company that provides such work will be sought again and again. Businesses simply cannot succeed without trust, and without integrity and objectivity, there can be no trust.

As Aerospace CEO, I was responsible for making sure that the company worked vigilantly to preserve the safety and security of the nation by delivering innovative, thoroughly researched engineering analyses in an objective manner with the utmost integrity.

In the space business, many factors have the potential to affect the design and construction of complex engineering systems, including cost, requested capability, geopolitical instability, and local politics. These factors can and do influence the design and engineering of space systems, but it is vitally important that the engineers designing and constructing these systems, and the companies that employ those engineers, never lose site of the task at hand and the serious realities of a given engineering challenge. The rocket must launch; the satellite must deliver its capability. It is Aerospace's objectivity and unbiased opinion that has always established its credibility and allowed customers to trust its recommendations.

This objectivity is sometimes difficult to maintain without its companion value, integrity—especially in those times where it might be more convenient to compromise than to maintain objectivity. In Aerospace's line of work, designers and engineers can sometimes be placed in the awkward position of being asked to do something for a customer that is very high risk, lacks technological maturity, or is beyond present manufacturing capability. In these situations, it's the responsibility of Aerospace employees to exercise good judgment, and when necessary, to say no. It's said that in the restaurant and retail industries, the customer is always right; however, when national space security is on the line, it's *data* that determines right and wrong. Customers, along with engineers, must recognize that there is always risk.

Integrity means that you don't back off from your ideas or recommendations just because they aren't welcomed. You do the research, you honor your values, and you make decisions based on what you know is best for your mission and your customers in the long term. It's a hard thing to do, but it's a lot easier when you have a strong, well-established set of values guiding your every move. The corporate leader is the steward of those values, infusing them into the company's culture so that they can guide the work of every team member at every level of the organization.

VISION

While values determine the underlying identity of a company, its vision enables us to determine where we are headed. A vision needs to be inspirational to the people within the organization, and provide stability and guidance during times of uncertainty and hardship. It also needs to be a bit more flexible than a company's values, so that relevance can be maintained in times of industry upheaval and evolution.

Until recently, Aerospace's vision was to be the federally funded research and development center (FFRDC)* for military space work in the United States. Essentially, the single goal of the company, and its founding purpose, was to assist the U.S. Air Force by providing thorough and unbiased mission assurance work and technical expertise for all of the Air Force's space-based missions. Today, that vision has evolved to include support to programs and customers far beyond the U.S. Department of Defense. Aerospace is still the FFRDC for the Air Force's national security space programs, but in order to remain relevant in a changing industry, the company needed to diversify its work and engage with new partners in both the international and commercial sectors.

When I started in the space business, there were really only two nations that could put things in space and had access to space data—the United States and the Soviet Union. As of the writing of this book, there are more than 60 countries that now have their own national

---

* FFRDCs are nonprofit organizations that support government science, engineering, and technology development. FFRDCs do not compete with industry and do not manufacture products, eliminating conflict of interest. FFRDCs operate as strategic partners with their sponsoring government agencies to ensure the highest levels of objectivity and technical excellence.

space assets, and the list is growing. Aerospace is no longer in a position where focusing solely on one government customer is technically effective or economically sustainable. By diversifying its clients, Aerospace has been able to take advantage of all of the incredible growth that is taking place in the space market today. Diversification is essential to maintaining the relevance of any company. Relevance is determined through continual self-reevaluation and, if needed, evolution. Aerospace is in the process of its own evolution—one that will allow the company to better serve both its military and civilian customers by embracing a broader understanding of the use of space, the operations of space-based systems, and the applications of space technology.

At Aerospace, we've had to reshape our initial corporate vision into something broader and more inclusive as a result of the expansion of the space industry. As CEO, it was my responsibility to oversee that process and help redefine the corporate vision to reflect new goals and opportunities. In 2015, Aerospace's official corporate vision statement became "To provide engineering solutions to the nation's most complex challenges." These words are more encompassing than Aerospace's past vision statements, and allow employees and customers to know that Aerospace isn't simply a space company, but a dynamic, multifaceted hub of engineering and technical excellence. This more expansive vision has guided many recent corporate actions and helped the company focus on new opportunities in addition to the essential FFRDC work Aerospace has always performed for the nation.

One such corporate action was the creation in 2015 of a new business unit called Vaeros, which is focused entirely on leveraging the company's expertise in new markets. For years, the Aerospace name served, unintentionally, as a barrier between the company and non-space clients because of its specificity. It didn't reflect all of the depth and diversity of Aerospace's services. In 2012, my leadership team and I realized that we had to address that very issue, and Vaeros was our solution. The new business unit now serves as an effective extension of the Aerospace brand, continuing to work in the national interest.

It's absolutely critical that you, as a leader, examine your company in relation to its competitors and its industry at large. When you feel your company has fallen behind, it may already be too late to make meaningful change. It's important to always absorb and map the context surrounding your company. By observing significant trends in business and technology, you can make effective leadership decisions that can increase the relevance, value, and efficiency of your organization.

As CEO, I periodically evaluated Aerospace to make sure that the company's goals and corporate mission put everyone in the best position to succeed in both the short and long term. Many times I was required to be the voice of reason, steadying the ship so that we maintained the course we charted many years earlier. Other times I was required to make significant changes that resulted in the permanent evolution of the organization. The corporate vision should always play a significant role in the navigation process and, though it shouldn't be altered very often, it should be repositioned and reinvigorated when necessary to accommodate the evolution of an industry and the sustainable growth of a company.

STRATEGY

It's essential to have overarching goals for your company. While a strong set of values and a clear vision can drive the culture of your business and allow employees to feel invested in your mission, they produce results only when they are carefully implemented as part of a focused strategic plan. Let's take chess as an example. In that game, you don't simply improvise your way to victory. You study the board, study your opponent, and develop a strategy based on the potential outcomes and risk factors.

At Aerospace, I conducted two week-long strategic planning sessions per year for the leadership team, which included my direct reports, the general manager for corporate communications, and the general manager for human resources. We would publish our strategic plan every two years and measure the success of programs against the plan every six months. We didn't just make goals and promises, we also devised implementation plans with intermediate milestones, measuring our progress and holding ourselves accountable for achieving our milestones.

During these strategy weeks, my team and I would discuss numerous issues affecting the company and work to develop plans that could position the company to succeed given the current state of the global marketplace. These sessions were about examining the bigger picture and how the company fit into the broader landscape of the space and engineering business. These weeks were significantly different from our day-to-day operations, because the purpose was to examine current practices and to decide which needed to be terminated, which needed additional investment, and which needed a course correction. We also used the time for tours of customer locations and team-building activities.

Team-building was a useful side benefit of these strategy sessions. By working together as a group, you have the opportunity to gain insight and understanding of each other's perspectives and concerns. This builds trust and respect, and enables people to work together effectively during challenging times. It also provides an environment where people can learn to disagree with ideas that may be suggested without being disagreeable. When different ideas and backgrounds are given the space to coexist, great plans for the future can be developed.

As a company of engineers and scientists, we were typically comfortably embedded in the details of our customer's most critical projects. We liked facts, data, and problem-solving in a very focused way. During strategy week, the corporate leaders were forced to take the opposite approach. They had to look at the big picture, evaluate the corporate vision, and come up with initiatives to ensure that the vision was reflected in all of the company's work.

President Dwight D. Eisenhower once said, "Plans are worthless, but planning is everything." He was making the point that the process of planning results in learning, even though the final plan might be short-lived. The strategy must be developed and implemented in order to effect change. Economies change. Industries change. Technologies change. The value of a strategy or plan, without fail, erodes with time. As a result, we need to keep our strategies agile and use feedback to recalibrate them to accommodate our changing world.

EXECUTION

Having vision, values, and a strategy are essential to figuring out what to do, but the hard work comes when you actually turn those ideas into meaningful action. Leaders are uniquely responsible for the implementation and execution of any plan. Without proper attention to execution, goals cannot be reached. Leaders need to communicate effectively and they must have a vision, but they must also be able to produce results that customers will value—otherwise their organization will cease to exist.

Throughout my career, I was responsible for the execution of numerous plans and strategies. One significant example took place in 2011, when my team and I implemented a plan to take on more private sector business to meet shifting demand and government budgetary pressure. Aside from the new expertise and infrastructure we needed to build internally, we also needed to communicate effectively with our employees and customers about recent paradigm shifts in the space business.

In 2011, the U.S. government space enterprise was beginning to leverage partnerships with the commercial space market. At the same time, the major defense contractors were recognizing that the government space business base was decreasing and the commercial business base was increasing.

Aerospace had always been almost exclusively focused on supporting the government's space needs. It was now essential for the corporation to embrace the changes occurring in the marketplace and expand the scope of its business to be inclusive of, and adapt to, the increasingly diverse makeup of the industry. To make it work, my team and I implemented a plan to translate our valuable expertise and knowledge to address the current problem of architecting responsive and resilient space systems. We had to do this while continuing to deliver the support and expertise needed to ensure mission success for both programs already in the development cycle, and those legacy programs already on orbit. This required developing a communications plan to help existing employees understand their role in this new world order. It also required developing tools to enable them to deliver the valuable support to existing programs in a more cost-effective manner.

Aerospace's two primary customers, the National Reconnaissance Office and the United States Air Force, are an integral part of the U.S. national security mission, and many Aerospace employees join the company intending to devote their skills and abilities to that mission. After September 11, 2001, that mission began to evolve even while the pressure on the national budget remained. Our role as the trusted technical advisor for space increased in importance at the same time that we needed to expand our investments in cyber, innovation, and system of systems architecting.

One significant difference for the company at this time was the fact that throughout its history, it had taken pride in the fact that the work it did received little public recognition. The work was a pure act of service to the nation. Such a modest approach can become detrimental; if no one knows your company exists, they will tend to ask, 'why are you necessary?' In the 21st century, organizations like Aerospace were routinely required to justify their existence to the government. The sense of national pride still remained in these organizations, but strategic communication was becoming a necessity. Although Aerospace's stellar technical contributions had not diminished, it became important for senior customers and elected officials to understand and appreciate the work that was accomplished by the company.

This resulted in the development of an annual value letter to senior customers, which provided highlights of the great accomplishments and contributions to their mission during the previous fiscal year. Technical execution was no longer enough. We now had to execute both on an engineering level and on a public-relations level. It was my job to make sure we were covered in both of these areas, and I worked tirelessly to make sure that we saw our strategic plans through—from a technical and communications standpoint.

As discussed in Bossidy and Charam's *Execution: The Discipline of Getting Things Done*, execution brings precision and discipline to the strategy. It enables you to methodically accomplish your bold goals one step at a time. Your technical success is ultimately what customers will reward. That success is a direct result of your ability to execute a plan—from start to finish. As the leader of the organization, you are responsible for providing the drum beat and being the driving force behind the execution.

A critical part of being that driving force is to incentivize execution and performance. When you observe an employee succeeding in the execution of his or her role in the larger plan, it is very important to shine a light on that effort so that everyone in the company can see it. A word of thanks and appreciation can be highly motivating to both the employee being praised, and those others hearing that praise. Letting your employees know when they do something well or execute properly, in my experience, is just as important as holding an employee accountable when something goes wrong. Great execution breeds more great execution, and you want to continue to reinforce that positive feedback loop whenever you can.

The bottom line is that planning and preparation are crucial, but without execution, your company doesn't really have much of a product to offer its customers. As a leader, you must do everything in your power to make sure that execution throughout your organization is at the highest level. Plans may not be everything, but results sure are.

In this chapter, we've touched on the key leadership responsibilities of leading an organization: values, vision, strategy, and execution. The leader needs the entire team to engage in developing these elements in order to deliver exquisite results, but it is the leader who will determine if the organization can be more than the sum of its parts and achieve great things.

Values, vision, and strategy may feel like abstract concepts, but they deserve a great deal of attention. It's very easy to think about strategic initiatives without consulting your values or corporate vision. Yet, as leaders, that kind of shortsighted decisionmaking will create missed opportunities. Leadership choices must be well-considered and intentional, and decisions must have substance and depth behind them. By aligning your strategic initiatives with your corporate values and vision, you can make cohesive, effective leadership choices that will greatly benefit your company, your employees, and your customers.

CHAPTER TWO

Leading through Unexpected, Uncertain, and Intentional Change

CHANGE IS A FACT OF LIFE AND
A FACT OF LEADERSHIP.

On June 17, 2014, my day started out like any other. I went to the office, attended meetings, and met with members of my leadership team. It was a hectic morning, and I recall dealing with the typical slew of phone calls and urgent emails that regularly punctuated my days. After a quick lunch—standing, cold sandwich in hand, speakerphone switched on—I left the Aerospace campus in El Segundo to attend a meeting in downtown Los Angeles. The meeting went well, and afterward I returned to my car with my head spinning, full of ideas and action items for the rest of the day.

As I approached the Aerospace offices, I began to notice a change in the familiar streets surrounding our campus. Cars were stopped along the road. People on the sidewalk were on their cell phones, pointing animatedly down the road. Traffic was being redirected. Police officers and squad cars were everywhere. All of this unusual activity seemed to be emanating from the Los Angeles Air Force Base, located directly across from, and attached by a pedestrian bridge to, The Aerospace Corporation.

Clearly something big was going on. It had to be a security threat. No other action would require road closures and instigate a police response of this magnitude. Once this reality set in, I focused my thoughts. I immediately asked myself how the situation might be affecting our company and the type of response we would need in order to mitigate a potential threat to the safety of our personnel:

- Is security engaged?
- What communication lines are established?
- Have we established contact with the Air Force base, the El Segundo Police Department, and military and civilian personnel?
- Do we need to lock down the connecting bridge between Aerospace and the base?
- What's the threat and how does it affect our people?

Dozens of other questions raced through my mind as I began making calls to my leadership team on site. I quickly found out that our security team was already directly engaged in monitoring the situation at the base. A potential active shooter had been sighted entering the base, and police and Air Force personnel were working to track down the suspect. My emergency response team immediately snapped into action. We had already locked down our entire campus to protect our employees, so the next step was to bring together our communications and security teams to deliver a series of messages to our staff, to inform them of the situation and to keep everyone coordinated and prepared for potential action. In addition, I established communications with the air base group commander and throughout the incident I received constant updates on the efforts of their security team.

The lockdown lasted for four hours. Fortunately, it turned out that the potential suspect had been mistakenly identified as a shooter, and the threat was not credible. Gates were reopened. Our employees completed their workday and left for home. Cars began to flow again down the once-empty streets. Normalcy had been reestablished, but the unexpected events of the day left a lasting impression on me. I was proud of our response. We stayed calm, protected our most valued assets (our employees), communicated effectively, and put ourselves in a position to work through any outcome that might have resulted from this uniquely unnerving situation. We were prepared in the face of the unexpected.

By definition, unexpected events occur without warning and rarely at a convenient time. These events result in a leader having to make important decisions without the luxury of a lengthy period for analysis. As a leader, your job in these instances is to rapidly build situational awareness, so you can respond effectively to the challenge. As CEO, I was faced with unexpected situations on an all-too-frequent basis. To support my decisionmaking, I always relied on my past experiences and on the contingency planning my team and I prepared and exercised throughout the year.

All kinds of things happen in the world every day. As a leader, you need to always be aware of current events—even if they seem irrelevant to your business. Given the right circumstances and participants, a remote-sounding news headline might just affect you and your company in ways you could never have imagined. For instance, if a deadly bombing occurred somewhere in the world, I learned to verify the location of all of our traveling employees to ensure they were in a safe place. Fortunately, a situation never arose during my tenure where we had anyone directly in harm's way, but employees and their families always appreciated our efforts to offer assistance.

The successes and setbacks of other organizations in the space business also frequently impacted our own operations at Aerospace. For example, when there was a launch failure for a project that Aerospace was not involved in, we still would stop and take notice. In these difficult situations, my immediate reaction was to feel a strong sense of empathy because I knew that some group of people in my industry was having a really challenging day on the job. More importantly, however, even though such a failure wasn't my direct responsibility, I knew that everyone at Aerospace would want to understand the root cause of the failure, to avoid making the same mistakes our peers had, and to gain valuable knowledge for future missions. Failed launches, though incredibly unfortunate, could still present valuable learning opportunities for everyone in the industry. Every organization is connected—and in an industry where a failure can mean the loss of billions of dollars, every failure and every success matters.

What I've learned in my years as a leader is that while it's not always obvious how you or your company might be connected to an event taking place in our increasingly global society, it is important to expect the unexpected and to take note of these events—you need to analyze them and you need to learn from them. As a leader, it's your job to understand inter-connections and be prepared for unintended consequences.

Being prepared for the unexpected means that you need to have pre-established responses to potential scenarios. Of course, there are millions of possible scenarios that can impact your company and there's simply no way of preparing for all of them. Leaders must examine what's probable and identify those scenarios that have a reasonably high chance of occurring. The more probable a situation with high consequences is, the more time you should spend on preparation and the development of processes that can automatically kick in when needed. Making investments to ensure these responses are well thought out and practiced will pay dividends over time.

Throughout the year, Aerospace organized exercises to prepare for natural and man-made disasters. Our employees assumed different roles during event simulations and we learned things about our processes that we could never have anticipated. Of course, actual disasters are far more intense, complicated, and emotional than our staged exercises were, but the muscle-memory response that practice creates allowed us to stack the deck in our favor. At the very least, we had a shared experience to lean on in case a real crisis did come our way.

I also made it a point to learn from the responses of other organizations that have dealt with significant crises. I compiled articles about various leadership responses, engaged in discussions with my team, and when possible, I would even try to reach out to the leaders at those affected organizations to see what lessons they had learned. As a CEO, you can never be over-prepared for a crisis situation. The livelihood, and the lives, of your employees can be put in jeopardy by any number of unexpected actions. When you learn through observation, preparation, and planning, you put your organization in the best possible position to mitigate threatening situations and weather the metaphorical, and sometimes literal, unexpected storm.

THE UNEXPECTED ORGANIZATION CRISIS: GOVERNMENT SHUTDOWN

There were times as CEO when I was tested in extraordinary ways—where unique circumstances required me to call upon all of my preparation and planning to make difficult decisions in a limited amount of time. There is no better example of this than the events that unfolded at the beginning of Aerospace's fiscal year in October 2013.

Our company had spent the previous 18 months getting ready to sign a new contract with the Air Force. That contract accounted for most of our future business revenue and represented the lifeblood of our company. In September of 2013, my team and I believed that we had done everything we could to get a contract in place in a timely manner. We had an agreement on the terms and conditions, and it seemed like we were in a perfect position to lock in a very strong and mutually beneficial contract.

Unfortunately, there were other forces at play that would have a profound effect on our seemingly successful negotiations. The U.S. Congress was engaged in a highly contentious debate about the federal budget. My staff and I were following the federal budget negotiations closely, but we expected that differences would be resolved and our politicians would eventually come to a budget agreement. It seemed unreasonable to expect that our Congressional representatives would choose to shut down the federal government rather than pass a new budget... but that's exactly what happened.

The ripple effects of the government shutdown were rapid and impactful, moving outward from the epicenter in Washington, D.C., across the nation, and to the rest of the world. It was a clear demonstration that no matter how isolated or unrelated to each other people and organizations might seem, we are all interdependent. The negative impacts of the shutdown were felt by millions of people, and those impacts varied greatly across the U.S. Everyone experienced the shutdown differently. At Aerospace, our experience was very direct and very significant.

On the evening of the shutdown, I received notification from the Air Force that everything we had discussed and agreed upon in our new contract was going to be placed on hold. Suddenly, the smooth transition that we had worked so diligently to manage over the past 18 months had been transformed into chaos.

We had never prepared for an event quite like this, which meant we had to learn on the fly and rapidly build situational awareness. We didn't have even a full day to plan our response—we had to act right away. That's when my leadership autopilot kicked in. My first order of business was to gather my leadership team and start collecting information about this rapidly evolving situation. I wanted to know how our employees were going to be affected, how our contracts were going to be affected, and how things were progressing in the nation's capital—the only place where a solution truly existed.

The CEO is the only one inherently responsible for seeing the complete 360-degree view of an unexpected situation. To form this view, you need to examine an issue from multiple perspectives and engage with affected parties to understand the full scope of the situation. Your established relationships and networks, formal and informal, become essential to the quality and speed of building your situational awareness. Through them you can gather the data and information that you can use to start piecing together the bigger picture.

With the government shutdown, we were acutely aware of what I call "first moment" opportunities—those early moments where key decisions are made that affect everything that follows. As with a stroke, the first 60 minutes can make all the difference in determining the eventual health of a patient...or a corporation. If you continually observe the unfolding context of a crisis situation, you may find that certain windows of opportunity are open temporarily. You need to take advantage of any beneficial opportunity you can find during those first moments. Procrastination or delay can result in a missed opportunity or irreparable damage. Decisive action can provide you with a chance to succeed despite immensely challenging circumstances. You have to engage immediately.

Once you *are* engaged, you will start making a series of cascading decisions about what to do next. In a crisis context like the government shutdown, our decisionmaking was largely responsive to the actions of our customers and our elected representatives. We were receiving new data and information about the shutdown on an hourly basis. My team and I examined the data, handed out or executed assignments, and reported back to each other about what we'd learned. Communications materials of all kinds needed to be delivered to our customers and employees, and we made sure to distribute that information as quickly as possible. Our jobs and our livelihoods were on the line and everyone deserved to be informed of any progress that was being made.

Early on in this process I had to make some very difficult decisions about our employees. Our Air Force customer made it clear that only our most essential operational activities could continue in the face of the shutdown. All personnel not engaged in such activities would be sent home without pay for as long as the political stalemate lasted. On an emotional level, it was devastating for me to have to inform many of our hard-working and incredibly dedicated employees that they couldn't work during this time. These were people who were absolutely devoted to our mission, with families to provide for and remarkable projects to complete. Approximately 80 percent of our workforce was impacted during the first week of the shutdown. During those initial days, I remember walking onto our once-bustling campus and seeing thousands of empty parking spaces. The halls were quiet. The cafeteria was closed. It was surreal.

CEOs feel pain. We feel sadness. We feel emotion and, despite what many people will tell you, we are human. One of the most frustrating aspects of the government shutdown experience was the fact that I couldn't look our employees in the eyes and tell them exactly when they could return to Aerospace. That decision was beyond my control. What *was* in my control, and the best way to honor our employees, was for me to communicate regularly with them throughout the shutdown and to fight for their jobs each and every day. I contacted anyone and everyone who could help to resolve the situation. The leadership team worked tirelessly to find any bit of leverage we could use to reopen Aerospace at full capacity. The situation was very dynamic and the rules of engagement were fluid, but each day we made progress and after three days, things started to turn around.

I relied heavily on the advice of our board of trustees, our senior customers, our senior leaders, and my numerous external contacts in the industry. By the end of the first week of the shutdown, we were able to get a large number of our employees back to work. By the end of the second week, the shutdown had ended and everyone was back in their offices—once again supporting national security.

During this entire ordeal, I never forgot our values. In fact, those values informed many of my decisions and choices. I kept thinking about the commitment we make to our people, and I knew that this was an opportunity where we needed to work incredibly hard to make sure that we honored that commitment. We couldn't prevent the shutdown from happening, but we could work hard to support our employees—and through them, our customers—and I believe we did that in a very meaningful way.

An unexpected organizational crisis of this nature doesn't happen frequently, but it can tell you a lot about your abilities as a leader. There's an old saying that goes, "the same boiling water that softens potatoes, hardens eggs," meaning that intense circumstances can reveal what we are made of on a fundamental level. In extreme situations, great leadership hardens like the egg and poor leadership softens and breaks apart like the potato. With a strong team to lean on, as well as preparation, experience, values, and vision, you can strengthen your abilities as a leader and meet challenges with a confidence that inspires and, most importantly, serves the irreplaceable people that comprise your organization.

Uncertainty

No matter how much you anticipate and prepare for unexpected events, uncertainty is a constant companion of all leaders. Vagueness and variability are key elements of every leadership situation.

In 2014, when speaking about the future of the U.S. military, Defense Secretary Chuck Hagel said, "The biggest challenge is the uncertainty of the budgets."[†] He was right. The budget, in any business, is always a key factor in making decisions, and budget uncertainty is a challenge that most leaders face. A breakdown in the budgeting process is what led to sequestration and the government shutdown in 2013 and, in general, most organizations struggle with long-term planning—particularly budget planning—in an uncertain environment.

Whether you are funded by Congress, as we were, or driven by unpredictable and volatile revenue streams, budget uncertainty is very common. The Aerospace fiscal year runs from October 1 to September 30. We rarely know in the beginning of our fiscal year what our budget is and how much business we'll do that year. Nevertheless, we have to

---

[†] http://www.defenseone.com/management/2015/01/hagel-budget-uncertainty-biggest-challenge-facing-military/103065/

start working on October 1. The budget is going to come, it's not unexpected, but each fall there's a fair amount of uncertainty surrounding when the budget will be firmly set and how the budget will change from the previous year. As a result, we start the year in October with a best-guess for the budget, knowing it will continue to evolve until the third or fourth quarter.

If the actual budget is larger than our best guess, we can adjust staffing plans and modify schedules to fully respond to customer requests. If the actual budget is consistent with the budget from the previous year, we can maintain the operating plan from the just-completed fiscal year. A third scenario involves a budget that is slightly lower than the year before. In response to a small budget drop, we can work to make our operation a bit leaner without affecting our employee base. Certain temporary austerity measures can come into play that allow us to meet the budget requirements without enduring major staffing disruptions. Finally, a fourth scenario, in which we experience a large budget cut, can occur. This option comes with the highest level of uncertainty and unpredictability in terms of outcomes. It also presents us with our greatest challenges in terms of managing business, personnel, and morale.

If a severely constrained budget is delivered to your company, you have to assess and identify what it is you are going to *stop* doing. You have to communicate the impact to your existing customers, and you have to figure out how to pay for the costs associated with stopping work and the liabilities associated with a reduction in force. These are very unpleasant realities, and the required actions are often challenging and painful. As CEO, you must make the challenging decisions that will result in the reduction of your services and workforce. These choices won't make you popular or beloved, but they are the kinds of decisions that leaders must make. It's your job to remain focused on the future of the organization regardless of the circumstances. You will adjust your resources accordingly and communicate clearly with your entire organization about the nature of the budget reduction and the adjustments it has necessitated.

Uncertainty, at its core, is an odds game and as a leader, you are required to make calculated decisions based on the likelihood of certain outcomes taking place. You cannot control the future, but you sure can plan for it. There's no excuse for being blindsided by uncertainty. Leaders must see the big picture while simultaneously reading between the lines and in so doing, recognize the potential the future holds for your organization.

INTENTIONAL CHANGE

Of course, not all of the change we experience in business involves uncertainty or an unexpected crisis. In fact, strong leadership necessitates a large degree of intentional change. Every member of your organization should be engaged in a self-regulated process of continuous improvement. A new idea can come from any level within the organization, and your employees must feel empowered to demonstrate leadership and make innovative contributions in every situation, even if it requires the reevaluation of successful and/or time-tested procedures. Your organizational culture should support outside-the-box thinking and champion intelligent evolution, not fear-driven status quo.

As CEO, your support can lend a significant boost to the successful execution of new ideas. When people see that you encourage new ideas, resulting in a new concept that you think will benefit the corporation, they're more likely to prioritize it, knowing that they have a powerful ally to help them see it through. A CEO has a great deal of responsibility to make sure that good ideas are supported and given the proper opportunity to flourish. The inherent power of the position can be used as an effective instrument for making deliberate changes in your organization. By simply listening to and supporting your employees and their ideas, you can steward the effective changes needed to remain relevant and successful on a commercial and cultural level.

Internal Change (Organizational Change)

One example of intentional change during my time at Aerospace involved the development and standardization of an internal core technical process that our company used (and continues to use) to identify whether a space program is ready to proceed to its next major phase, or milestone. This core process originally started as a group of checklists generated by subject-matter experts. For the most part, the old process worked, but as the annual launch rate increased from two or three per year to eight or ten per year, the disparities from program to program became apparent. We needed a better and more consistent process. So we worked diligently to develop a new review process that included a cross-functional group of internal experts who met to identify the myriad important aspects of these complex space systems that needed to be reviewed as part of our pre-launch preparations.

We started by codifying all of the critical actions and processes that are essential to a successful launch. We intentionally carved out time to study and learn from the teams that had been the most successful in the past. Over a period of years, we evaluated best practices, collected new input, retested our methods, and refined, refined, refined. We ended up building a very streamlined and efficient review process that we've come to trust and rely upon, known as The Aerospace President's Review, or APR. The APR has come to be regarded as one of our company's crown jewels and, in many respects, it helped to revolutionize the work we do and reinforce the confidence our customers have in us.

Another example of intentional change was in our launch verification process. Even though the national security space community enjoyed an unprecedented record of 100-percent launch success from 2000 to 2015, there were many critics of the launch mission assurance process, who claimed that it took too long and was too expensive to be cost-effective. Further, while this impressive record of launch success yielded a robust on-orbit capability for communications, navigation, and remote sensing, the marketplace was changing as private investments in commercial space grew. The emergence of new launch vehicles like SpaceX's Falcon and United Launch Alliance's Vulcan provided new options for delivering national security space payloads to space.

Our launch verification process needed to evolve to leverage the new ways of doing business that were being developed by the commercial entities, so as to reduce the cost to the government in maintaining its national security space capabilities. The challenge was to alter our well-tested process to understand and assess these new approaches without losing the trust of our customers and without increasing the risk of a mission failure. It was challenging for us to evolve, but over the course of two years, the company made great strides in improving what was once considered an untouchable recipe for launch success.

To identify and lead intentional changes involving important core processes, you have to start with a very simple notion: that every part of your organization, no matter how successful, can always improve. You also need the ability to understand and appreciate the powerful forces of context and culture. Both elements can make or break a change effort. A change in context—for example, in our case, new entrants into our industry—can alter the significance of a previously reliable solution. It's not enough to say, "Well, it worked last time, so it should work this time." That mentality is limiting and completely unacceptable. The fact that something worked in the past tells you almost nothing about how effective or affordable it will be in the future. As the leader, you need to keep an eye on context at all times and constantly question your practices, even if those practices have brought you great success.

I've often repeated the old adage to my team that the three things you can't avoid in life are death, taxes, and change. Death happens once, taxes happen once a year, but change happens all of the time. You need to constantly evolve to make change work for you. The successful leader gets in front of change, whether it is intentional, unexpected, or uncertain.

As a CEO, keep in mind that your company can't be navigated like a nimble car that immediately responds to your decisions. You're steering a large ship saturated with its own culture and momentum, and it takes great care and forethought to control. The ship can't be turned abruptly; it must be guided gently. Discussions must take place. Information must be delivered. Consensus and commitment must be established and solidified. As a leader, you may push for the adoption of a new concept and the response may be far more cautious and skeptical than you would like. A leader has to be flexible in presenting and achieving new ideas and directions, and be ready to educate the employee population when confronted by skepticism. Leadership takes time and space. It takes transparency. It takes sustained effort if you want to produce amazing results.

CHAPTER THREE

Decisionmaking

The CEO role comes with a great deal of prestige, but that prestige is tempered by an incredible amount of responsibility. When things are going well, a CEO will get most of the credit; when things go poorly, a CEO will get most of the blame. This is true regardless of the type of work the CEO's company does; when you're the boss, you have the ultimate responsibility for all the company's decisions and actions.

Making sound, levelheaded choices for your company on a consistent basis is one of the CEO's primary responsibilities. Decisionmaking is a skill that grows stronger with practice. The higher you go in an organization, the more practice you'll get in making increasingly complex decisions that affect increasingly large groups of stakeholders. You will also find, as more challenges compete for your limited resources and attention, that you have less time than ever before to both contemplate the issues requiring your decisions, and to weigh the consequences. In order to manage this dizzying workload, you must become an incredibly efficient decisionmaker.

NO MAGIC CURTAIN

Throughout your career you may have assumed that people at higher levels of your organization have more insight than you, better perspective, extra data, or some magical recipe of all three to help them make perfect decisions. This usually isn't the case. Decisionmaking, at every level, almost always involves a healthy dose of uncertainty. No one has all the answers—not even the CEO.

When I became CEO, it finally hit home—there was never going to be a magic curtain to pull back where I would find all the answers. There was never going to be a wizard or oracle for me to consult who could provide me with omniscience. In fact, I found that, as CEO, many of my employees and customers were looking at *me* as that omniscient wizard! It had become my responsibility to make far-reaching decisions and advise staff members throughout my company, even though I knew that I didn't have all the answers. So it was incumbent upon me to hone my decisionmaking process, rely on my team for additional knowledge and expertise, and to make tough choices without a safety net.

CEOs are responsible for the efforts of their company, yet they often cannot effect necessary changes on their own; they need the participation of a web of stakeholders and advisors. The CEO can't just make things happen with the snap of her fingers. It often takes much more than a single perspective to arrive at wise decisions concerning an important and complex issue. It is also important to recognize that *not* making a decision is a decision in itself. Inaction is always an option, but it also comes with its own costs and consequences.

Decisionmaking occupied a tremendous portion of my day-to-day work as CEO. For me, a typical day might include decisions on technical risk assessment, succession planning, board of trustees interactions, long-range corporate business strategy, unexpected realtime crises, research and innovation, employee benefits issues, corporate social responsibility, and administrative decisions that keep the doors open and the lights on. Each decision I made came with its own context, costs, timetable, and risk, all of which informed my approach and, in turn, the choices I made. I was not simply multitasking, but actually recalibrating my brain each time a new decision was called for. I learned that as CEO, it was absolutely essential to develop the ability to shift gears quickly and focus on multiple decisions across a wide spectrum of subjects in a short amount of time. Where other parts of the job felt like either sprints or marathons, this type of rapid-fire decisionmaking was more akin to a decathlon—where a host of different skills were tested in challenging ways.

THE TYRANNY OF THE URGENT

It's simply impossible to focus on all aspects of your business simultaneously, 100 percent of the time. Even the most brilliant individuals have their limitations, and part of being an effective leader is understanding what needs to be prioritized and what can be delayed or delegated to another team member. Every issue needs to be addressed with the appropriate level of focus and attention, but there are significant limitations to what you, the leader, can personally achieve on a daily basis. This is what I often refer to as wrestling with the "tyranny of the urgent." As a leader, you are constantly presented with seemingly urgent issues by your team members, but you have a need to balance those fast-moving concerns with the work that you know is important and essential. There are always going to be "urgent" meetings that can fill your calendar and occupy your time; it's your responsibility to make sure that you spend enough time every week doing the critical and essential work that only you can do.

Decisionmaking, by its very nature, is a double-edged sword. No matter what choice you make, there is another choice (or series of choices) that will go unrealized. In most instances, your decisions will make some people happy yet disappoint others. Accepting that your choices will rarely be met with universal acclaim is critical to accepting the responsibility of leadership. You don't become a leader to please everyone all of the time; you become a leader to make the tough decisions that bring the greatest net benefit to your organization. These decisions are rarely universally accepted, nor should you expect them to be. Sorting through different viewpoints and making bold, educated decisions is your job and it's inherently polarizing. Again, your job isn't to make everyone happy. It's to make everyone better off, and there's a big difference between the two.

RULE OF LEAST REGRET

For me, there are two key skills that have significantly contributed to my decisionmaking process: learning to see the big picture in any situation and balancing the short-term and long-term impacts of any choice I make. When you improve your understanding of the big picture, you can more accurately assess and predict the ramifications of your decisions. It's a mindset similar to that of a chess player. In chess, you are assessing potential moves and playing the game based on probability, preparation, and past experience. In leadership, you want to think deeply about the numerous potential ramifications of your decisions and make the best possible choice you can. You need to think about how your moves and countermoves will fit together and how they will change as the result of developments in the outside world that are beyond your control. You need to learn when to attack, when to retreat, and when to bide your time and wait for new opportunities.

It's incredibly important to take the time to intentionally balance both the short- and long-term effects of any decision. What seems like a good quick-fix or short-term solution can become the source of a long-term problem. For example, I was in the position, on a number of occasions, to decide between leasing or buying/building additional office space for Aerospace. Under certain conditions, leasing was the right choice. In other situations, building a permanent facility was more cost-effective and strategically beneficial over time. In all of these instances, my eventual decisions were always based on the overall impact to the company and our customers in the short-term and the long-term.

When faced with a decision, a key question I would often ask myself was: "What happens if I'm wrong?" It's surprisingly difficult to come to terms with the potential results of failure. No one wants to be wrong. No one even wants to *think* about it, but as a leader, it's part of my job. To account for the potential fallout from an incorrect decision, I developed a policy called the "rule of least regret." Essentially, this means that once you've considered all of the potential negative outcomes of a series of choices, you should aim to select the choices, or sequence of choices, that minimize avoidable risk. I found that in many instances, the best way to reduce risk was to make incremental decisions rather than sweeping ones.

If you break down a decision into a series of implementable steps, you give yourself the time to gain valuable perspective. You can take an initial action, and once it's been implemented, you can take a step back and observe how information or situations have changed. You can then adjust your approach to accommodate the results of that initial action and any unexpected changes it has caused. This multistep process allows you the flexibility to adapt to and maintain relevance in these new contexts. As a result, you can better minimize risk and maximize effectiveness for your organization.

ORGANIZATIONAL DECISION: REDUCTION IN FORCE (RIF)

One of the toughest decisions I had to make during my tenure as CEO was the decision to implement a reduction in force (RIF), or employee layoff, in 2012. Although layoffs are not uncommon in the aerospace and defense industry, The Aerospace Corporation had always prided itself on its exceptional workforce stability. Unfortunately, in the mid-2000s, the U.S. government made significant reductions to its defense budget—Aerospace's primary source of revenue—and in turn, impacted Aerospace's ability to maintain its current workforce. With a new, leaner budget in place, our existing staff levels were too high and needed to be reduced so that our company could continue to provide support in the years ahead. There were no temporary measures that could alleviate our new financial limitations. Difficult decisions needed to be made.

Any decision that involves the involuntary departure of valuable employees and the attendant negative impacts to their families is challenging and emotionally charged. In 2012, I had to make some gut-wrenching choices that would result in the disruption of the lives and financial security of people whom I knew personally. There wasn't a single part of me that wanted to make the decision to reduce the Aerospace workforce, but as CEO, it was my responsibility to do just that in order to secure the future success of the company.

Furthermore, the social context for the reduction in the Aerospace workforce this time was much different than it had been for RIFs in the '80s, prior to my tenure as CEO. Retaliatory cyber threats and viruses were becoming more common and incidents of workplace violence had changed the way Aerospace, and the rest of the world, looked at large-scale layoffs. We weren't just dealing with an emotional blow to our culture and our workforce, but a potential security threat as well.

As a result, the security of our people and information systems had to be considered while designing the RIF implementation plan. The Aerospace leadership and I eventually made the difficult decision to escort terminated employees out of the facility immediately after they were notified of their layoff, to minimize the likelihood of a security incident. Understandably, this upset many of our employees. It upset me. We were treating loyal employees as potential threats instead of the beloved friends and colleagues we knew them to be. However, it was the right decision.

It was the right decision because I knew that we simply couldn't afford to ignore the very real security threats that existed. I asked myself, what are all the options? What would be my regret in choosing each of these options? How do I minimize that regret? What if someone brings a gun to the office? What if someone plants a debilitating virus in our computer system? Is it worth the risk? After considering all these questions, the choice was simple: we were going to protect all of our employees and we were going to do so by hiring security to assist with the layoff process. This decision was a difficult one, because hiring extra security could be seen as an insult to the departing employees, who had served the corporation faithfully and well. Ultimately, though, it was the right decision—it was my responsibility to ensure the safety of the remaining employees and the security of the corporation as a whole.

It's a painful truth, but in today's world sometimes we must choose security over civility. There are dangers that exist now that simply weren't there in the past. We need to adjust and adapt, even if the results are, at times, less than ideal. In my mind, the safety of all my employees outweighed everything, and I made the choice that best protected them from harm. It wasn't a choice that I enjoyed making, but it was the choice that needed to be made.

TECHNICAL CRISES: A STRUCTURED APPROACH

The emergence of an unexpected technical crisis can provide yet another challenging context for decisionmaking. In the space business, an on-orbit anomaly that damages or disables a space system can create a very serious, and time-sensitive, challenge for both leadership and technical staff. In these situations, everyone involved needs to carefully think through the entire decisionmaking process so as to limit mistakes and develop solutions to a rapidly evolving set of problems. Things change quickly in space and sometimes it's impossible to fully restore a damaged system, so everyone needs to be willing to obey the rule of least regret and make challenging, timely decisions.

When Aerospace is tasked with mitigating on-orbit anomalies, a very formal process is intentionally set in motion to address the problem in a systematic and unbiased way. As CEO, it was my job to make sure that the process operated smoothly and that my staff was in the best possible position to succeed. I always wanted to monitor their technical progress without interrupting or adding unnecessary pressure to their efforts, so I made sure to respect their boundaries at all times and only selectively interject my opinions and critiques in realtime. They were the experts and they needed to be treated as such. Micromanaging would be disruptive, and disrupting our process was the last thing I wanted to do.

It was also my responsibility to ensure that we prepared our staff members to deliver difficult messages to the customer, so that we could make certain our perspective would be heard regardless of the outcome of the anomaly analysis. In the space industry, very expensive, government-funded assets can be compromised by even the slightest malfunction, and the customer is acutely aware of the tremendous consequences this can have for both taxpayers and national security. At Aerospace, we always needed to be prepared to deliver honest and thoughtful messages to the customer that addressed the legitimate concerns of the public. We also needed to prioritize which of our customers and board members had to be informed of and/or invited to participate in the resolution of a given anomaly. I worked to make sure that we were in control of our messaging at all times, both internally and externally, and that we kept our customers, employees, and the public informed in a way that didn't compromise our technical efforts.

The overall process of handling an on-orbit anomaly at Aerospace was designed to minimize the influence of bias on the technical work we were doing. We always wanted to make decisions for the right reasons, without being influenced by personal or political agendas. As engineers, my Aerospace staff members were never really interested in assigning blame, only in finding solutions and fixing technical problems. As a leader it was my job to make sure that our company's priorities were always properly aligned and that my staff was never distracted by the potential repercussions of their work. I was there to absorb criticism. I was there to keep the focus on finding the path forward. I was there to preserve the scientific credibility of the work at all times and to not let agendas or schedules dictate our decisionmaking.

Regardless of my best efforts, however, bias—even for logic-focused engineers—is still very challenging to overcome. As human beings, we appear to be wired for preferences, pre-conceived biases, and assumptions. Biases affect everything that we do, so the work of a company like Aerospace, which is designed to serve as an unbiased technical resource, can be incredibly challenging to protect from a person's inherent tendency to judge and influence.

I've seen many instances where, for whatever reason, a customer or employee has been personally driven to choose a specific solution or outcome before it has been scientifically validated. For a company like Aerospace, that should never happen, and a large part of my leadership responsibility was to make sure that the scientific process was never compromised by bias. We had to definitively prove our outcomes and analyses. We had to evaluate problems from multiple perspectives, not just the perspective that best accommodated our launch schedule or cost constraints. This commitment to avoid bias is how we built our credibility and maintained our value as a company.

INPUTS TO DECISIONMAKING

As a senior leader you'll often be faced with difficult choices that don't always have an obvious answer. When gathering data and opinions to help inform a decision, you may find yourself in a situation where two or more of your trusted advisors are advocating opposite positions for the decision at hand. Each advisor's position will have its good points, but you have to choose which one is best for the company overall. It's a difficult position to be in, but one that is common for corporate leaders. You'll need to make the hard choice, which might often be a third option that incorporates the strongest elements of both positions.

For example, I worked with my team to decide whether or not Aerospace should invest in purchasing or renting a new facility in Chantilly, Virginia. I had to make a choice between two seemingly "right" decisions. One of my advisors encouraged me to rent, given the tremendous upfront cost of a new construction project and the uncertainty surrounding the government budget—our primary source of income. Another advisor made the case that it was exactly the right time to build, since the construction market was slow, interest rates were low, and there was a uniquely suitable piece of land in Chantilly that was available for purchase. By gathering as much background information as possible, and relying on the expert opinions of my advisors, I was able to analyze the potential risks and benefits of each decision. Both advisors gave me accurate input and recommendations from their points of view, but ultimately I decided that the benefits of building outweighed the benefits of leasing. Later that year we started work on a new Chantilly facility and when it was completed in 2014, the new campus helped to solidify Aerospace's commitment to its East Coast customers and staff while, in the long term, reduce operating costs.

LEVERAGING YOUR EMOTIONAL INTELLIGENCE‡

Regardless of a leader's commitment to make unbiased decisions, it's still very easy for her to become emotionally attached to an issue that needs to be addressed. These emotional attachments can affect a leader's ability to be objective, see the big picture, weigh all the necessary options, and make a decision that, ultimately, does the most good. Your ability to assess the emotional landscape and identify potential emotional land mines is critical to identifying acceptable courses of action. Let's look at an example.

From 2000 to 2016, healthcare costs rose dramatically across the United States. The competitive healthcare packages that we had always offered our employees were becoming increasingly expensive to maintain. It was my responsibility to find a way to minimize costs while still offering a reasonably priced plan that did not overburden

---

‡ Emotional Intelligence: A type of social intelligence that involves the ability to monitor one's own and others' emotions, to discriminate among them, and to use the information to guide one's thinking and actions (Mayer and Salovey 1990).

employees and their families. It was a challenging position for the company to be in and I needed to find a way to balance my emotions with my leadership responsibilities.

The resulting decisionmaking process was all about finding ways to reach a middle ground where the employees and the company could both comfortably absorb a rate increase without suffering serious consequences. In this situation, my emotional inclination would have been to give our employees anything they needed when it came to health care, so that they could best serve themselves and their families. However, the cost to the company would have been prohibitive. Ideally, I would have done anything to support my fellow team members based on the assumption that it would make them better employees; I also recognized that the increase in employee cost-sharing would be a hardship for some. But I knew making a decision that was all in the employees' favor would have compromised my responsibility as CEO to protect the corporation as a whole.

In my leadership role, I wasn't just responsible for the health of my employees, but the health of the company. A strong company empowers its employees and allows them to thrive. The financial health of the company ultimately impacts the ability for its employees to maintain their jobs, and in turn, take care of themselves and their families. Everything is connected. So my decision on healthcare cost-sharing ultimately reflected that mutually dependent relationship between business and employee.

AFTER THE DECISION HAS BEEN MADE

Execution

Whenever you make a decision, you expect it to drive action in order to achieve results. As a leader, the fact that you have made a decision doesn't mean your work is done—not by a long shot. You also have to understand what's required to execute and support the decision afterward to ensure the desired outcome is achieved. This will likely require a considerable amount of effort and communication to publicize the decision and provide a rationale for why you believe it is the right choice. Building alignment and commitment around the decision takes skilled leadership. As CEO, you're working to execute from a broad organizational standpoint, as opposed to a divisional or individual standpoint. There is a lot of work that goes into coordinating the efforts of your team, your customers, and yourself in order to see that a decision is effectively executed. As you lay out your implementation plan, you'll need to create opportunities to build continuity and alignment. Decisionmaking is just the beginning.

Accountability

Leaders are accountable for their decisions in various ways. First and foremost, leaders need to know if their decisions are reasonable and can be successfully achieved by their team members. If a leader has built trust and credibility with her team, the team is likely to follow whatever direction she sets for them, often without any questions or contradictions. This type of faith is gratifying to a leader, but isn't always a good thing. Team members must know when to question a direction. They must feel empowered enough to share their viewpoints and fully engage in the decisionmaking process. It's a leader's responsibility to make sure that team members are involved and engaged, so that the entire organization can come together and work intelligently to execute effectively. Leaders need to take responsibility, not only for the decisions they make, but for the organizational environment that will eventually turn their plans into a reality. The entire organization needs to be positioned for success.

Even with the best intentions, things don't always work out as planned. Part of a leader's responsibility is to stay engaged and keep track of how things are going after a decision is made. If there are signs that your initial decision isn't producing positive results, you need to be proactive. Make changes to your plan or create a new path. Usually, the sooner you can detect, evaluate, and remedy a flawed decision, the better. Bad news rarely gets better with time, so take action and make needed adjustments.

Reconstruction: Justifying a Decision

As a leader, you will frequently be asked to make a decision without the volume or quality of data that you would like. Further, at some point after a decision is made, you may be called upon to justify it. Sometimes it will be revealed that you made the wrong choice. It's part of the job. You will face criticism, but you will also be able to learn from your mistake.

I've found it very helpful to go back and review a decision after the fact, reconstructing it, with full understanding of what we knew and when we knew it. This process of self-examination helps me to understand what information was missed and whether or not it was available at the time of the decision. During my career, this kind of after-the-fact analysis allowed me to evaluate everything from hiring decisions to the way our team handled on-orbit satellite anomalies. I may not have always made the right decision the first time, but if I was able to figure out where we went wrong, then I could make sure it wouldn't happen again. Leadership isn't about being right all of the time—it's about learning from your mistakes.

Decisionmaking is a critical leadership responsibility. It allows you to shape the future of your organization, make improvements to infrastructure, and adjust to new contexts and business climates. A CEO will receive credit when decisions work out, and blame when they don't. Again, that's just part of the job. All decisions have consequences, but the consequences of the decisions you make as CEO have greater potential impact. In order to mitigate risks, it's important to build a strong organizational culture that emphasizes integrity and empowers team members to voice their opinions and share their expertise. As CEO, you will bear the brunt of the responsibility for all decisions, but you can always make better choices when you have a strong set of values and an effective team to rely on.

CHAPTER FOUR

Stacking the Deck

THE TACTICS OF STRATEGIC LEADERSHIP.

To this point, I've made it clear that leadership isn't easy. It's not a skill that is learned overnight and there are no real shortcuts to success. But there are ways to prepare, to give yourself an advantage so that you can work as efficiently and effectively as possible, regardless of the challenges you may face. I refer to this preparative process as "stacking the deck." In cards, stacking the deck is a euphemism for cheating, but in leadership it means to optimize every aspect of the workday that you can—from the design of your calendar to the selection of travel options, and even to adjusting your personal sleep and dietary rituals. There are only 24 hours in each day, and to be an effective leader, you need to figure out how to make the most of every minute.

As CEO, your time and attention are split between a large number of people and issues, which makes it essential that you develop techniques to help you organize your schedule, prioritize your challenges, and perform at a high level throughout the day. You need a process to determine which responsibilities you will personally manage, which ones you will reject, and which ones to delegate to others. As professionals, our impulse is always to do all the work ourselves. Leaders, however, can't afford to be this generous with their time. Leaders need to exercise discipline and personally manage the essential responsibilities that will have the greatest impact on their organization and employees.

When I became CEO, I was amazed by the number of tasks, the diversity of the subjects, and the speed with which I had to engage, decide, and act on each topic. Time management classes that I had taken throughout my career provided me with some basic organizational tools, but in no way captured the intensity and mental gymnastics that were required to get through a typical day as CEO. I learned through experience—both successes and failures—and more than anything, I learned to listen to myself and respect my limitations while emphasizing my personal strengths.

A CEO'S CALENDAR

For me, the calendar was the core document that defined each one of my days as CEO. When properly constructed, a calendar can be a framework that enables success—but when it gets out of control, it can be an impediment to efficiency. I once analyzed my calendar and found that I averaged more than *100* meetings per month. Some meetings were scheduled for only 15 minutes, while others lasted as long as two days. An average day had six to eight meetings, but oftentimes I was faced with days that were completely full—topping out at 13 meetings in a single workday, as this example shows:

Time	With	Subject of Meeting
7:30 a.m.	Aerospace General Counsel	Review of legal issues
8 a.m.	Technical Editor	Discussion about upcoming publication
9 a.m.	Security Managers	Review of results from recent security inspection
9:30 a.m.	Launch Program Personnel	Pre-brief for launch program
10 a.m.	Engineering Staff	Formal engineering review
11 a.m.	Customer Group	Classified meeting to review a risk reduction proposal
12:15 p.m.	Visiting VIP	Lunchtime discussion
1 p.m.	Aerospace General Manager	Discussion of feedback on recent review of an on-orbit anomaly
2:30 p.m.	VP Physical Sciences Laboratory	Briefing about an upgrade to Aerospace laboratory operations
3:30 p.m.	Corporate Communications Team	Discussion about my upcoming keynote address at an industry symposium

Time	With	Subject of Meeting
4:30 p.m.	Aerospace Early Career Employee	One-on-one mentoring meeting
5 p.m.	My Executive Assistant	Review of miscellaneous items and calendar
7:30 p.m.	Private University	Keynote speech at STEM event

On that day (and many others), there were so many people to work with and so many subjects to master—all contained within a single work day. The sheer amount of time that a CEO spends in meetings is staggering and can be overwhelming if poorly managed. There is no time in a busy schedule for inefficient meetings, and it is the responsibility of all the participants to come prepared and ready to contribute. As CEO, your job is to know and define your role in any meeting, so that you can deliver key messages to your team and help facilitate effective decisionmaking. Preparation is everything, and there is no excuse for not doing your homework on an important subject.

Given the wide variety of topics on my calendar, I found it quite helpful to set aside time at the end of each day to prepare for the next day's meetings. During this time, I would read proposals, do limited fact-checking, and even call some of the meeting participants to get background on the subject matter. I always found it immensely helpful when meeting attendees would send material ahead of time. This allowed all participants to get on the same page and, in my case, familiarize myself with the topic that I would be expected to weigh in on decisively.

The extra time I devoted to meeting preparation always allowed me to stay grounded and move smoothly between subjects, even on those challenging 13-meeting days, which pushed me to my mental and physical limits. At the end of those long days, I would still take the time to prepare thoroughly for the next string of meetings. It was a vicious cycle, but an effective one that improved productivity for everyone involved. On evenings before a particularly challenging day, I would often think of the famous saying, "Give me six hours to chop down a tree and I will spend the first four sharpening the axe." As CEO, you always need to be sharpening your axe, preparing yourself for the deluge of demands and questions that will face you each and every day.

DAILY RHYTHMS

It's important to remember that leaders—CEOs included—are human beings with very human limitations. They eat, drink, sleep, and breathe just like everyone else. They are not immune to fatigue and, quite often, the fast pace of their work schedule makes them more prone to reach their physical and psychological breaking points than non-leaders.

A CEO must learn to become as efficient as possible while still giving attention to your physical and mental health. Recognizing your own limits is an incredibly important skill to develop. People can easily be stretched too thin by work demands, personal responsibilities, or some combination of the two, and it's important to recognize the warning signs. Your health, your family commitments, and your essential downtime must all be tended to while still meeting the demands of the job. It sounds like an impossible task, and there are times when it is. There will be moments where you have to sacrifice some element of your delicate work/life balance, but you need to work to ensure those situations are not the norm. The body and the mind of a CEO are her instruments, and the organization cannot afford for those instruments to fall out of tune.

When you are not at your best you make mistakes, and as CEO, your errors can have huge implications. A weary CEO can't just skip speeches or meetings she doesn't feel like attending. She needs to be present and involved all the time. During my tenure as CEO, I did everything I could to balance my work, home, and personal responsibilities, to create an environment that allowed me to give every task and interaction my very best shot.

MAKING THE SCHEDULE

A great executive assistant is absolutely critical to the success of any leader in the corporate world. For me, that assistant was Larcine Gantner, and it was her wisdom, organizational expertise, and adaptability that allowed her to give me exquisite support no matter what challenges we faced. Her work was essential to my successful management of the corporation. It seemed that each day Larcine's responsibilities would change as a result of a new request or impending deadline, and she always amazed me with her ability to learn and calmly manage a schedule that was constantly in flux.

Larcine and I developed many pragmatic approaches to handling my schedule; one of those was to organize meetings based on the priority of the issues being addressed, positioning them to ensure that they would actually take place. We learned to schedule important meetings early in the morning. In our experience, 7 a.m. meetings had about a 99-percent likelihood of happening, so the most significant meetings were typically scheduled for that time. A meeting at 10 a.m. had about a 50-percent chance of happening as scheduled, and the later we got into the day, the higher the odds were that unscheduled events or delays would bump other items off the calendar. We referred to events that could potentially supersede preplanned meetings as "priority interrupts." These might come in the form of an unexpected visit from a key stakeholder, or a space system anomaly that required all hands on deck. Each situation came with its own unique context and usually required a judgment call from me—which would almost always then result in schedule changes.

A CEO is incredibly busy during the work week, but there is also quite a bit of work that falls on weekends and days off. Oftentimes dinners, community events, and award celebrations are held on the weekends, all of which serve to extend the CEO's work week. As CEO, you really are on call 24 hours a day, seven days a week.

Social events are also important, because engaging in casual networking with associates can be very valuable to a CEO and the overall image of a company. Awards ceremonies offer a particularly significant opportunity to show support for the achievements of colleagues and coworkers. Appearances at community events allow a CEO to send a strong message about corporate social responsibility and the values of the company. Aligning yourself with strong causes both inside and outside of the work environment helps to solidify your reputation and credibility, and that of the corporation. By showing up at these weekend and off-day events, CEOs can "walk the talk" and make it clear that they support great people and great causes.

In a few instances at Aerospace where we had an unresolved technical issue prior to a launch, I would meet with my team on a Saturday. This minimized distractions and ensured that everyone with a critical role in the effort would be present. At these meetings it was very important that each team member showed up focused and prepared, with the necessary support materials to make critical decisions. No one likes the idea of working on a weekend, but when it was necessary, everyone always rose to the occasion and treated those days like any other work day.

We all want to push our teams to achieve as much as possible, but it's simply unsustainable to bring your coworkers in to work every weekend. Not only will it make you very unpopular with your staff, but people can easily be overworked to the point of inefficiency. If your staff is physically and mentally worn down, mistakes will be made and performance will suffer. They need time to rest and recuperate in order to be at their best. So it's important to be judicious about defining the situations that require extra work hours and the ones that don't. I was always selective about scheduling meetings on the weekend, and on the rare occasions where it was done, I found that, in hindsight, the team agreed that our productive results justified the sacrifice of personal time.

I also used weekends to prepare for the week to come. These extra hours helped me establish broad situational awareness and stay up to date on the issues affecting all parts of the company. My direct reports would send an email to me at the end of each week and I would review them on Sundays. These short highlight reports were compiled by their managers and they served three functions:

1. To identify what had happened during the past week and list major events expected in the upcoming week
2. To highlight what the managers believed the senior leadership needed to know
3. To prepare me and them for "accidental meetings" with senior leaders and members of industry

The weekly reports also served as part of a "sensing and information" network that kept me and my team informed. When I examined these summaries from each group, I learned about new events, activities, and patterns that I had previously missed. Furthermore, I was able to see the connections between different departments and events, which allowed me to make better, more forward-thinking decisions regarding complex issues. The reports helped to ensure that I was seeing the bigger picture.

TRAVEL

Business travel can be incredibly beneficial and enable you to take part in critical work interactions away from headquarters; at the same time, business travel is very time-consuming. As with everything in the CEO toolkit, business travel must be used judiciously, so that it maximizes the benefits while minimizing negative impacts.

Traveling to a customer's location can make a world of difference. At the conclusion of many such meetings, I often had customers thank me for taking the time to visit with them in person. My presence showed that I valued them as a customer and that their project was being given the full attention of the CEO. These meetings were always beneficial because they allowed me to engage with the customer, communicate effectively, and analyze firsthand the risks that a given project might be facing. They also served to help build and strengthen important customer-client relationships that would be tested by future challenges or disagreements.

Moreover, often a customer would applaud or criticize an individual member of my team—providing me with a frank performance analysis that I wouldn't get in other situations. I could use that feedback to make decisions about personnel and future assignments. This sort of openness couldn't be replicated, and those direct interactions proved to be incredibly beneficial throughout my time as CEO. Business travel made those breakthrough moments possible.

By putting time in to meet with a customer and strengthen our relationship, I was always able to deliver difficult messages and speak "truth to power" when it came to critical issues affecting their projects. Our personal bonds allowed us to more effectively find solutions and mitigate problems. These relationships were always strengthened by face-to-face contact.

The entire travel process itself, however, can also be very exhausting and time-consuming. I am continually surprised by people who view business travel as a job perk. I've often heard people say "Gee, you are so lucky that you get to travel!" I always respond politely, but I'm thinking that their image of business travel does not reflect reality. Business travel is not a vacation. There are no beaches, relaxing massages, or luxurious nights out on the town. Just work, sleep, and airports—lots of airports.

Business travel can have a significant impact on your efficiency. In addition to the time lost while traveling from one location to another, most people are not at their best when they are sleeping away from home, missing out on family time, eating the wrong foods, disrupting their exercise routines, and suffering from endless workdays—all while combating the disorienting effects of jet lag. I have found, though, that with a little effort, it is possible to counterbalance some of the adverse effects and manage the travel challenges.

In addition to the typical things that frequent travelers do—boarding early, packing only what you can carry on, and getting preferred seats—I found that it was also incredibly helpful to check a flight's history when arranging my itinerary. If a flight has a poor record for on-time departure, it's prudent to reschedule. Similarly, I never booked the last flight out of a city unless it was an absolute necessity. If the last flight is cancelled, you will be stuck overnight in a city that you don't need to be in, which will have tremendous impacts on your calendar.

Finally, I discovered the benefit of adding an extra day of travel onto the front end of critical trips. I used to think arriving for a business meeting on the same day the meeting took place would optimize my time. I found, however, that on those days I would be tired, rushed, and at the mercy of flight delays and cancellations. Arriving the day before allowed me to be rested, acclimated to the new time zone, and ready for the day's challenges.

KNOWING YOURSELF

Whether traveling or in the office, it's absolutely necessary for leaders to take time to rest, recharge, and reduce stress levels. Exercise can help a lot in this area. Taking a yoga class or going on a long run is not only good for the body, but it can stimulate the brain and get new ideas flowing. Furthermore, casual discussion and even phone calls with friends or mentors can be mentally invigorating in a way that typical office collaboration is not. My own favorite tool for mental relaxation was incredibly simple, but surprisingly, I often had to work hard to build it into my schedule—a quiet time for thinking and planning ahead. It was a challenge, but I found that I was usually able to carve out a little bit of quiet time early in the morning before my first scheduled meeting or while I was on board flights for business travel. It was always immensely helpful to devote those small moments of downtime to thought and reflection.

It seems funny that CEOs need to schedule seemingly intuitive activities like "reflection" or "quiet time," but believe me, if it's not on the schedule, it simply doesn't happen. You need to make time for yourself. Each person operates differently and benefits from different methods of relaxation and regeneration. The key is to find what works for you and to make time for it. As a leader, you need to defend your mental and physical health at all times so that your performance level is always high.

Of course, there is always that nagging urge to get assignments done as quickly as possible. Deadlines are important and must be met, but sometimes leaders need to ask themselves and their teams whether or not a certain task must be completed right away. Sometimes the answer is yes, and in those cases you do your best with the limited energy you and your team have to offer. However, if the answer is no, the best course can often be to pause and circle back to that task after gathering more data and taking time for reflection. We may not like to admit it, but as human beings our attention and capacity to focus are limited and exhaustible resources. That's just a fact.

Sometimes leading means knowing when a team may actually benefit from a chance to take a break, recharge, and return to the assignment later on. It's not always possible or practical to take a step back from work, but when it is, you, as the CEO, may be the only person who can make the decision to give yourself or the team a breather. Sometimes the space and perspective a well-timed pause can offer makes all the difference in terms of the success of a project or assignment. It also sets a good example for the rest of the team. If they see you running yourself into the ground, they will assume that you expect them to do the same.

Stacking the deck means finding and using approaches that not only support you in doing your best each day, but will sustain and strengthen you over time. People often ask me my views on the balance between work and personal affairs in my own life. They ask me how I could manage such a busy schedule and still have some semblance of a life. These are tricky questions to answer because each day as CEO was different for me, and in truth, sometimes it was very hard for me to achieve a state of balance. In my experience, the frequent imbalance of work and personal affairs is simply a hazard of the CEO role, but it is a hazard that can be kept in check with proper preparation and planning. That's what stacking the deck is all about—being prepared to be at your best at all times.

CHAPTER FIVE

Complexity and Leadership

EVERYTHING AFFECTS EVERYTHING ELSE
—ADAPT TO LEAD.

SEEING THE TIP OF THE ICEBERG—BECOMING NO. 6

Prior to becoming CEO, I had 29 years to watch the senior leaders of The Aerospace Corporation tackle a wide range of problems. Although I joined the company after the retirement of its founder and first president Ivan Getting, his influence remained strong and pervasive throughout my career. His leadership helped define the high-quality standards and stellar technical reputation that Aerospace enjoys to this day. The CEOs who led the company after him each made their own signature contributions to the organizational structure and technical focus of the corporation.

Aerospace's second CEO, Eberhardt "Eb" Rechtin, introduced the importance of systems engineering and system architecting of complex space systems to the company. Rechtin also had the foresight to see that the changing demographics of the nation and the workforce would affect the corporation.

The third CEO, Sam Tennant, dealt with the challenge of the nation's response to the Challenger launch failure and the resulting changes in the launch vehicle architecture, an evolving customer base, and the need to develop the Aerospace brand beyond the Air Force commander for space and missile systems.

The fourth CEO, Edward "Pete" Aldridge, worked to convince the community, as he had done earlier as Secretary of the Air Force, that we needed a robust expendable launch capability in addition to the space shuttle. He also revalidated the need for the unique role of the FFRDC.

Fifth was William F. Ballhaus Jr., who took the helm directly after a series of launch failures and refocused the corporate effort on mission assurance.

I was number six.

I took careful note of and learned from the many valuable lessons my predecessors left behind. Eb taught me the value of engaging and strengthening the workforce. Sam was masterful at solving hard technical problems. Pete demonstrated the importance of building strong relationships across the customer community. Bill was brilliant at asking penetrating questions, which revealed the root cause of a problem. Each leader faced a different set of technical, business, and talent-related challenges that had to be carefully managed. Each was successful in his own way, and from my vantage point, they made it look easy.

On January 1, 2008, I became President and CEO of The Aerospace Corporation. Although I had been able to study and learn from the first five CEOs, I couldn't fully appreciate the magnitude of the complexity embodied in the job itself. I quickly learned that although I had watched carefully and prepared for my new position, I had previously only seen the tip of the iceberg.

When I reflect back upon my early days as CEO, the skill that I relied upon the most was learning agility. I was working to keep pace with challenges that I had never faced, questions from customers that I had never been asked, and serious business decisions that I was never before in a position to make. It was undoubtedly very exciting to be serving in such a dynamic position, but it was also intimidating. I learned very quickly that in order to be a successful CEO, I didn't need to be fearless and I didn't need to have all the answers to every question—I needed to be organized, decisive, and to the maximum extent possible, transparent. I found that with proper preparation and a dedication to quality in every facet of the business, I could be successful.

The complexity of the CEO role, more than anything else, is what made the job so challenging early on and it's also what made the job so invigorating for me during the latter part of my tenure. Typically, complex situations can cause a sense of paralysis or hesitancy on the part of a decisionmaker. In an effort not to make a challenging situation worse, there is a temptation to do nothing at all. But CEOs must transcend the impulse for inertia. Even when every member of your team seems comfortable with their indecision, as the leader, it is your job to break the complex problem into smaller, less complex, and solvable pieces. In this chapter, I will review some of the overarching concepts that account for the tremendous complexity organizational leaders face, as well as a few strategies I've developed for managing that complexity.

WHAT THE RESEARCH SHOWS (AND WHY IT MATTERS)

When you are the leader of an organization, you regularly deal with rapidly moving events, a significant amount of change, and a healthy dose of the unexpected. As a result, the scope of your vision needs to expand greatly to successfully manage all of the multifaceted responsibilities that will come your way. To keep up, you'll need to learn constantly and find ways of quickly building situational awareness to accommodate a very dynamic environment.

In the space business, systems are made up of millions of pieces and keeping track of each one can be a huge challenge. We're not talking about Legos here; these are expensive, technically complex components that require a tremendous amount of care and attention. In the space industry, complexity results from the sheer magnitude of a project's component parts combined with the harsh environment in which those parts have to operate. This kind of complexity is what John

Sterman calls "combinatorial complexity"—lots of pieces that can interconnect in lots of ways. Sterman also defines a second kind of complexity, one that is even more challenging for leaders to manage: "dynamic complexity"—complexity that arises from the way component parts fit together to drive dynamic and unpredictable system behavior.

Unlike combinatorial complexity, dynamic complexity can occur from just a few connected parts providing positive and negative feedback to one another. The resulting feedback loops are a key source of the unpredictable behavior that defines dynamic complexity.

In the real world, both types of complexity work off of one another and converge to produce an environment that is immensely complicated and fraught with interconnections and nonlinear repercussions. Systems thinking[§] and system dynamics[**] modeling are processes I've used to make order out of what feels like chaos and to provide some sense of control to my organization and myself.

UNDERSTANDING COMPLEXITY: IT WON'T GO AWAY AND IT CAN'T BE IGNORED

Complexity, and its component parts, can explain a lot about the behavior of organizations. The people that an organizational leader guides are, collectively, a "complex adaptive system (CAS)[††]"; the leader is trying to provide direction that results in alignment and commitment. In order to lead effectively, leaders must understand how such a system—made up of people—works. When leaders recognize the characteristics and dynamics of a CAS, they can bring out the best in colleagues, reduce conflict, and boost engagement to achieve goals.

[§] Systems Thinking: The process of understanding how those things which may be regarded as systems influence one another within a complete entity, or larger system (Senge, 1990).

[**] System Dynamics: A methodology for analyzing complex problems that develop or persist over time (Austin, 1988).

[††] Complex Adaptive System: An entity consisting of many diverse, autonomous components that are interrelated, interdependent, and interconnected, and act as a unified whole in learning and adjusting to changes in the environment.

Today's daily challenges are better managed when teams and networks are part of a system-wide net, gathering and sharing data, ideas, and constructive feedback.

One of the lessons of leadership in a complex environment is that context matters. You are operating in an environment that is connected, congested, contested, and usually overly constrained. If you change the context, then what worked in the past may not work as well in the present. Similarly, in a business, the processes, approaches, and even the products that once defined a company may become irrelevant in the blink of an eye. Can you imagine a successful American phone booth manufacturer in the 21st century? The answer is a resounding no, and that's because of context. There is little to no use for phone booths in an age of smart phones and global wireless service.

As a leader, it's vital to routinely assess the context in which you are operating and recognize that customer requirements can change. In order to do that, you need to create an environment where you and your team can ask challenging questions of each other. Questions like: Is this still the best way? Will that process be as relevant going forward? Is there a new technology that may work better? Is there a way to simplify the problem? Your approach to problem solving has to be nonlinear. Your path toward leadership success is, as Sheryl Sandberg in her book *Lean In* states, more like climbing a jungle gym and less like climbing a ladder. You must account for a multitude of outcomes and constantly be prepared to adjust and recalibrate.

In business today, you simply cannot rest on yesterday's success. You must be prepared for change, but oftentimes this concept of adaptability directly contradicts human nature. If something worked in the past, we want to keep doing it. People are understandably proud of their successes, and they should be. It's hard to reset and step away from something that was once the right answer. It's hard to embrace change. Sometimes just asking, "Should we reevaluate this process?" is seen as threatening to the culture of an organization. Yet as leaders, that is exactly what we must do—over and over again. Our metaphorical phone-booth company must transform itself into something relevant, or risk complete extinction.

The good news is that we, and our CAS, can learn and adapt to our evolving environment—all it takes is awareness and an organizational culture that promotes constant questioning and self-examination. If a team is willing to challenge the status quo, that team will undoubtedly be more effective, and most importantly, relevant. Relevance is predicated upon a strong understanding of context. If you don't know where you are, who your competitors are, and what your customers demand, you are on the path to becoming irrelevant. As leaders, we must constantly work to be effective in complex environments that change rapidly and swiftly transform time-tested procedures and products into cobwebbed relics of a different time.

THE NEED FOR SPEED (OR THE LACK THEREOF)

I can assure you that while the speed of change is certainly a challenge, it's not the whole story. In fact, if that's all you focus on, you'll miss some really important things. Don't let short deadlines or the threat of competition fool you; it is absolutely possible to move too quickly. A singular focus on speed always comes at the detriment of quality. Doing something rapidly can be very impressive, but doing something well and delivering it in a timely manner will keep you in business.

We live in a time when globalization, accelerating technological progress, and a culture of instantaneous access have combined to create a business climate that is more intensely competitive than ever. Everyone wants to be first, but as a leader, you are responsible for not only making sure that your organization can keep up with competitors, but that it can remain focused on its goals and core values. I've learned that taking the time up front to understand the context of a situation before making a significant decision is absolutely essential to my process as a leader. It doesn't matter how demanding my schedule is, I always make sure to carve out some time to step back and evaluate the options and the potential outcomes of my decisions. In the moment, it can feel counterproductive or even wasteful, but it's not. In fact, in my experience, evaluating your options against your priorities leads to more effective decisions in a complex situation. With a strong understanding of context, you can decide what really matters, both for the short term and the long term.

Having said that, I'll also tell you that there are times when the speed of an event can require you to make a decision on a complex issue based on your gut. You don't have enough time to thoroughly weigh all inputs and outcomes, so you make use of your knowledge and intuition to guide your decisionmaking. In these situations, having a process for critical thinking and step-by-step problem solving is crucial.

I'm a mathematician and an engineer. Methodical problem-solving is ingrained in my brain and it guides and informs all of my work. As a mother and a wife, this can sometimes cause problems (my family vacations include itemized checklists and obsessively detailed agendas that would terrify most people), but in a work setting, my logical approach has been invaluable. In a fast-moving situation, I try to take things piece by piece, solving a problem as though it were a detailed equation—step by step. In the moment, each portion of the process might feel insignificant, but eventually those well-executed parts make for a complete, and correct, whole. For me, it's always about understanding and then deconstructing complexity—even when time is limited and quick decisions are required. A decision may seem incredibly complex when it is first presented, but once its underlying components are exposed, we can easily see what needs to be done first, attack the individual parts, and develop a working understanding of, and an appropriate solution to, the problem.

You don't need to be a mathematician to problem-solve. But you do need to understand that every problem can be broken down into smaller, more manageable pieces. As a CEO, you will need to be able to quickly identify those pieces in order to make sound decisions at the speed of need.

SEEING THE BIGGER PICTURE

Another challenge that senior leaders face is seeing and understanding the bigger picture across multiple organizations and communities. Sometimes constituents, including your customers and employees, have an expectation that as a leader, you must know everything that is going on. Sadly, that's not possible. Leaders need to be well-informed, but as I've stated earlier, they can't be omniscient. There's always more going on than a leader will know, and that's normal. Being aware that you have limits and potential blind spots forces you to view situations from a wider perspective and allows you to lead broadly, rather than myopically.

To see the bigger, more complex picture, you must be able to generate situational awareness for a variety of circumstances. As the leader, you are kind of like a fisherman, sitting on the edge of your boat, rocking on a stormy sea, continually casting nets out in order to collect fish, or, in your case, gather information and context for a given situation. Sometimes the nets come back full of beautiful fish (good information), sometimes they come back empty, but more often than not, those nets are filled with both beautiful fish (good information), dead fish (old information), and some form of oceanic garbage (wrong information). Furthermore, bigger nets, though able to collect more useful information, require us to spend more time analyzing that information to make sure that it is accurate and relevant. As a result, you need to balance your time and the resources available for data collection to mitigate the complexity of a given situation.

I've discovered that the process of making your nets bigger can often push you outside of your comfort zone. In these situations, self-awareness can really make a positive difference. Self-awareness is a key factor in bolstering what Daniel Goleman calls your "emotional intelligence" (EI), which is a lot different than a person's technical savvy and IQ. It's incredibly useful to be smart and technically brilliant, but it's not nearly enough to make you an effective leader. You need EI to understand yourself, your customers, and your employees. Remember, context is critical. As a leader, finding a "correct" solution is only a beginning. The way you implement that solution and communicate it to your team is equally important, and ultimately, what gives that solution the best chance to succeed.

To be effective, your leadership "nets" need to include people who think differently than you do. These nets should consist of colleagues, your contractors, your advisors, and your customers. These people will all have a lot of good advice to give, but, again, context is key: good advice isn't always useful advice, particularly when you are making decisions on behalf of an entire company. Everyone that comprises your nets is dedicated to an individual part of a large system that you are responsible for. They don't, and shouldn't, think about things the same way that you do. After all, they are doing their jobs, which, in most cases, are more focused and specific than yours. The problem is that these specific roles can sometimes act as a barrier to building a net-wide understanding of the needs and goals of the larger system.

Roles in a complex system, while necessary for efficiency, can often limit our vision and obscure the big picture of what the system, as a whole, is trying to do. Most roles are not designed to see that big picture, but for a CEO, the big picture is your primary concern. You need to understand many different viewpoints and make decisions that benefit the entire organization. Those decisions can come at the expense of others with different roles in your system. These people might disagree with you in order to preserve the status quo or to protect their individual areas of interest. So, you need to recognize that behavior and address it.

NEW ACTIONS AND NEW RISKS:
AN INSEPARABLE PAIR

In the real world, actions create ripple effects—some expected and some unexpected. Leaders need to work to understand the potential ripple effects of an action before the action occurs so as to mitigate that risk. As an example, let's think about my favorite sport, college football, and my alma mater, the University of Southern California. Imagine you're the coach of the USC Trojans and it's fourth down with ten seconds to go in the fourth quarter. You're on your opponent's one-yard line and you're down by three points. What do you do? Do you pass? Do you run the ball? Do you kick a field goal? Do you call a trick play?

These are all possible choices with a host of different potential outcomes. As coach, your responsibility is to analyze your team and your opponent's strengths and weaknesses. It's an imperfect science, but you make a call based on the information you have and you try to manage risk in a way that best suits your team. There is no "right" answer in this situation. You look at your team, you look at their history, you look at how the game has been going up until this point, and you make the best decision that you can for your team at that specific moment. You can't predict the exact end result of your decision—and that's ok! Leadership is about making the best difficult decision you can make to put your organization in a position to succeed. It's not about predicting the future, but recognizing that whatever the outcome, you must be prepared to make the next decision.

Whether you're the USC football coach, or a teacher in a classroom, or the CEO of a company, the goals and the approach are the same: prepare yourself, prepare your team, make decisions based on your goals and vision, make decisions based on the abilities of your team, and always be willing to adapt and change.

STAKEHOLDER INVESTMENT

One of the biggest jumps in complexity I noticed in my transition from senior vice president to CEO was the increase in the size of the stakeholder base. CEOs have to manage a portfolio of stakeholders that all require attention in different ways. Stakeholders include customers, board members, vendors, suppliers, former executives, employees, and former employees, among others. Every stakeholder group is unique and should be treated as such. One thing that all stakeholders have in common is that they want to be heard and know that the CEO values their perspective. It can be very challenging for a CEO to balance the needs of different stakeholders because each group has different goals and interests. Creating alignment among all of the different stakeholders is one of the most complex problems a CEO will face.

I was often presented with competing, yet equally valid, demands from different stakeholders. "The CEO Report"[‡‡] describes this as facing situations where you must choose between "right ... and right." In a complex world, simple right and wrong solutions don't appear nearly as often as we would like.

So leaders must deal with ambiguity. Sometimes there are multiple right answers to a given problem. In order to make an optimal choice, you have to establish guidelines for prioritizing resources, time, and attention to best serve the company, customers, and individual constituents. You also need to be adept at building consensus. If you can zoom out and think about the big picture, it's a lot easier to make these types of challenging, and seemingly ambiguous, decisions.

Stakeholders can provide numerous challenges for a CEO, but they are also a vital resource. They can be an extension of you and they can expand your knowledge base exponentially. You may be a CEO—heck, you may even be a rocket scientist—but you'll still never know

---

[‡‡] "The CEO Report." *Heidrick & Struggles*. Heidrick & Struggles and Saïd Business School at the University of Oxford, 21 Jan. 2015. Web. 1 Mar. 2015.

everything you need to know to do your job. In order to do that, you will need to frequently rely on the expertise of others. It is very important to invest time with your stakeholders. Stakeholders are your "sensors," and are key contributors to your leadership "net." They can tell you what is happening in the world around you. They will see emerging problems or things that you cannot detect. These people are influential because they have networks of their own that can be relied upon for support and guidance. If you know your stakeholders and understand their needs and values, you'll be able to rely on them and form great partnerships.

IT'S A MAD, MAD, MAD … AND COMPLEX … WORLD

The Economy

As we all know, the state of the economy has a huge effect on business opportunities and thus an enormous impact on the role of the CEO. No matter where you look today, everyone is dealing with limited resources and cost constraints. It doesn't matter what industry or specialty you're in or how critical your particular service may be, everyone is carefully considering how to spend, and sometimes save, their limited funds. Today's economic realities mean we're continually required to do more with less in an environment that is always demanding more from us. The state of the economy drives competition and it can require you to be innovative and develop multiple business models across your organization. One day you might be partnering with someone, and the next you're competing with them on a different project. It's a delicate balance, and as CEO, you always need to be aware of the impacts of the fragile state of our increasingly global economy.

The Community and General Public

The community and general public are part of the larger complex environment in which any company is embedded. What you do as an organization can affect the public, and vice-versa, so building external relationships can have significant consequences. Corporate social responsibility and good corporate citizenship are hallmarks of successful leadership. For instance, a leader may choose to make her company more environmentally friendly or partner with local schools to provide tutors or equipment as part of a company's larger effort to give back and pay it forward. These community-enhancing efforts are not insignificant parts of the job. They are complex leadership opportunities that will require real attention from a CEO, but they can establish genuine community loyalty and support for a company, which can come in handy during times of crisis and uncertainty.

The benefits of community outreach were apparent on numerous occasions throughout my tenure at Aerospace. One example was when we engaged in a round of talks regarding the Base Realignment and Closure (BRAC) process. BRAC is designed to analyze and make recommendations about military base locations across the U.S., occasionally resulting in the Department of Defense closing or moving military bases. BRAC outcomes can be very significant to Aerospace, because the company's headquarters is co-located with the Space and Missile Systems Center at Los Angeles Air Force Base. During our discussions, it was very helpful to create awareness among local, state, and federal elected representatives and local residents so they could provide support and highlight the potential impacts of moving our business to a new locale. The strong bonds we had built within our community gave us a great deal of support during the BRAC discussions and ultimately, the Air Force base remained open.

To summarize, the strategic leadership skills that are required in a complex and dynamic business environment are significantly different from those that would lead to success in a stable business environment. The risks and the threats in a complex world are inherently different and much more challenging. The complexity we see today is derived from the sheer number of issues that must be addressed simultaneously, as well as the interactions, interdependencies, and feedback loops that are rapidly changing the context for business. The technology that is currently available to leaders, their customers, and their competitors creates a need for the constant collection and analyses of data to discover the information that will support critical decisionmaking.

Complexity requires that we learn to excel in the face of change, the unknown, and the unexpected. Successful leadership requires that, first, you continually assess what is changing and what impacts these changes will have on your organization; second, you learn to distinguish between the things that can be controlled and the things that cannot, to avoid expending energy on things that are beyond your sphere of influence; and third, you develop strategic action plans that are agile and resilient. This three-part approach to managing complexity will significantly increase the likelihood of success in achieving your strategic vision.

CHAPTER SIX

Building Your Team

A SYNERGISTIC, HIGH-FUNCTIONING TEAM IS ESSENTIAL TO ACHIEVING YOUR GOALS.

Throughout this book I've discussed the many ways today's complex environment affects leaders and the way they lead. Leaders do not work in a vacuum or work alone. Global events influence every sector of business and leaders in this era must be willing to recognize their own limitations and remain agile and adaptable in times of change. An individual person can't know or understand everything. To be truly effective, a great leader must also have an excellent team to provide guidance, expertise, and perspective. In this chapter, when I refer to my team, I'm speaking primarily about my executive leadership team of 15 people, which included all of the senior vice presidents and vice presidents at The Aerospace Corporation, as well as my executive assistant.

It may seem obvious, but it's worth emphasizing: the stronger your leadership team is, the more effective your organization will be. Your team is an extension of you, the CEO. This team, in many respects, is leading the company. The team gives you the opportunity to communicate clearly and directly with representatives from every major area of your organization, to make sure everyone is aligned, coordinated, and committed to the same vision. Corporate decisions are made and implemented by this team. Because so much is required of the team members, and so much rides on their decisions, you must be absolutely certain to hire the right people for the job.

SHAPING THE TEAM

Some of the first assessments I made early on in my tenure as CEO involved evaluating job fit and establishing whether or not we had the right people in the right positions. In my case, having come through the ranks of the company, I knew most of the team members and I knew the unique skill sets that each one brought to his or her roles. I also had a full understanding of Aerospace's organizational culture—both its remarkable strengths and its weaknesses. In addition, I knew the customer base—knowledge gleaned over the course of decades with the company—and had some ideas for how Aerospace might anticipate and address those customers' future needs.

Regardless of whether you've come up through the ranks of a company or were hired externally, it's smart to strike a balance between protecting the existing corporate culture and pushing that culture forward, taking on new challenges with existing staff and recruiting fresh talent. Ideally, the goal is to keep the best parts of the company intact while improving and realigning the parts that need it. That balance between old and new is most pronounced, and most challenging, when it comes to hiring decisions.

Retirements and other causes of attrition can provide opportunities for change in a company's talent base, but fluctuations in the business environment can also precipitate the need for new talent. When vacancies occur on your team, they provide an opportunity to pause, assess the roles, consider them in the context of customer needs and the external environment, and determine what, if anything, needs to be modified. The significance of these opportunities for change is even more pronounced when they involve senior leaders.

Throughout my career at Aerospace, the culture had always been traditional and very conservative, yet when I became CEO, I knew it was my job to transform that culture into something more agile, innovative, and forward-thinking. My team and I needed to work to carefully manage risk instead of avoiding risk and maintaining the status quo. Each opportunity to bring on a new member of my leadership team was an opportunity for me to ask, "What is the right skill set for where this company is headed?" I wanted to bring people on board who were committed to the work the company was doing, but who still maintained enough perspective to make changes and improvements wherever necessary.

On occasion, when working to build a strong team, I was forced to make some difficult decisions about simultaneously redefining existing roles and developing new roles to address talent gaps. I learned quickly that the process of repositioning team members needed to be handled swiftly to minimize disruption.

On a broad scale, talent management requires the optimization of both the overall organizational structure and the individual positions within the organization. Sometimes that means making an untraditional choice to reinvigorate the system as a whole. When I was still new to the job of CEO, Aerospace's executive vice president retired and I decided not to immediately hire a replacement. The executive vice president had historically operated as a critical support system for the CEO, serving as a buffer between the CEO and his staff and customers. Since I was still building relationships and forming bonds, I wanted a more direct connection with my team, so I didn't fill the position immediately.

The ensuing months without an executive vice president gave my team the chance to learn what I expected and needed from them, and I, in turn, learned what they needed from me. This allowed us to connect and synchronize our working relationship so that, eventually, we intuitively understood one another. The direct connections that we built during this time were incredibly valuable. I learned that though a CEO may be busy, she must always make time to connect—in person—with her team. There's simply no other way to build an effective understanding of expectations and challenges.

In shaping the team, you must be willing to both introduce new concepts and strengthen previously established methods of operation. For example, early on in my tenure as CEO I recognized that Aerospace would benefit from a strong corporate communications department. In the past, the department had been internally focused and event-driven. I wanted to expand the accountabilities of the organization to include ownership of the corporate brand across all media platforms, both internally and externally, as we presented ourselves and our products to the outside world. I also wanted my leadership team to begin incorporating this outward-looking approach into their decisions and actions.

The expertise we needed to make this communications transformation was not available in-house, so we hired a strong external candidate to develop the capabilities of the rest of team. While her position was not a direct report to me, I empowered her from the beginning by bringing her into our senior leadership meetings, and I encouraged the rest of the team to draw on her experience in communications and media relations. I also clearly explained that this media focus was part of a long-term commitment the company was going to weave into its business practices—not a short-term project team members could opt into or not, at their discretion. With a new initiative like this, a key part of the leadership role was to ensure that my team members had the support and training they needed to be successful with their increased levels of engagement with external communication channels.

Within a year, the team was well-integrated, more collaborative, and aligned with our new communications approach. It was apparent to the board of trustees, to customers, and employees that the senior leadership team was speaking with a unified voice.

SETTING EXPECTATIONS

Each time that you step into a new leadership role, you will implement a new vision for how the team should operate. To bring about the changes you want, you must be clear in defining the vision and make sure that your team understands its role in bringing that vision to life. I set out to take my already high-performing team to the next level by establishing several expectations. From that point forward, they would be expected to:

1. Serve as exemplars of the corporate values and as stewards of the company
2. Build a strong identity with a common vision, a sense of ownership, a common operating picture, and a shared destiny
3. See the bigger picture and develop the external network
4. Maintain a tone of respect and cultivate diversity in teams and ideas
5. Develop themselves and their staffs, and to plan for succession

As I've already stated, you need to live your values every day if those values are to have any meaning or impact across your culture. Your people need to see the values driving your behavior, and the values need to be explicit. Whenever you can, you must link your values to the significance of the work you are doing. You must serve as a good steward by guarding the reputation, success, and well-being of the organization. If you can do all of that, your team is likely to follow your lead.

When working with my executive team, I always emphasized that if anyone on the team failed, we all failed together. Our fates were intertwined and our metrics, our performance-appraisal process, and our incentive plan all reinforced our collective responsibility. As a leader, your first responsibility is to your own organizational group, but I also expected all of our corporate leaders to always give attention to the challenges and goals of the company as a whole. I made this expectation very clear, and I followed up on it with both a performance-review process and a merit process that valued a strong commitment to the corporate mission. Managing an individual piece of the organization was just one requirement of my corporate leaders. They were given substantially more credit and recognition for reaching across the organization to collaborate and create opportunities for innovation beyond their minimum requirements.

For example, when Aerospace started the process of building a new facility in Chantilly, Virginia, the capital dollars needed for construction and initial operations had to be drawn from across the entire corporation. At the time, the biggest user of capital dollars in the company was the Engineering and Technology Group (ETG), and thus, ETG was primed to absorb the greatest financial impact from the project. Over the next few years, capital spending for ETG was cut in half. Obviously, this was a significant change to their budget and it required coordination with every organization to maintain customer satisfaction during the project buildout. Everyone needed to understand that the ETG reduction had a purpose and was in the best long-term interest of the company.

Once the new Chantilly campus project was complete, we restored the prior budget levels in ETG so that managers could work off their backlog of purchases. It was a four-year process that was executed without disruption, either internally or externally, because the entire team recognized that it was an opportunity to increase the value of the company. Instead of complaining about their budget reductions, the team members were helpful and pitched in wherever they could, confident that their short-term sacrifice would benefit the organization as a whole.

A key ingredient for achieving a unified, team-oriented culture within your organization is to continually emphasize the "big picture" ramifications of your policies and decisions, while simultaneously encourage systems thinking among all team members. Everyone needs to be on the same page and understand the goal they are working toward.

BUILDING NETWORKS

A powerful way a team can quickly begin to see the bigger picture is by building effective and targeted networks. I encouraged each member of my team to expand his or her functional networks outside of the company. This allowed specialists representing different parts of the company to connect with like-minded individuals throughout our industry and beyond. As these individual networks expanded, Aerospace gained access to an increasingly broad range of information and resources that could be called upon in times of need.

The reach, speed, flexibility, and diversity of networks makes them an invaluable asset to any organization. Networks, unlike teams, are fluid, ever-changing collections of people, talents, and personalities, and each one is entirely unique—no two networks are the same. Networks can be activated in realtime around issues of common interest. Gaining access to strong networks is an incredible skill, and it is directly reliant upon an individual's ability to find, develop, and foster genuine relationships.

Networks thrive on mutuality, authenticity, and value. It takes time and effort to build a quality network. I let all of my team members know that their networking would initially require a great deal of dedication, but would ultimately pay dividends. Over time, our networks gave us access to relevant and timely information on numerous occasions. By putting the hard work in up front and establishing new relationships, my team was able to build a series of interconnected networks that helped Aerospace through many challenging situations.

Once you extend your reach into new networks, you will encounter a large variety of diversity in people and ideas. Diversity is an incredibly powerful leadership tool that brings a strategic advantage and can provide vital solutions in complex environments. However, bringing together a diverse group with multiple points of view can also bring disagreement. As the leader, you can create an environment for healthy debate and disagreement within your team. By establishing a respectful and open environment, your team can discuss its differences in a way that encourages participation, not isolation or social discord.

On my team, members were encouraged to disagree with each other and share dissenting opinions, but only if they did so in a respectful manner. I asked them to challenge each other's ideas without attacking the holders of those ideas, and to engage in logical and thoughtful debate. I wanted to hear everyone's opinion and I wanted everyone to know that their individual opinion would get the attention it deserved. My goal was to encourage participation so that we could thoroughly analyze challenging situations and determine well-considered, effective solutions.

During meetings, the leader first needs to act as a moderator: watching, listening, and trying to see which team members are fully engaged and which ones are holding back. The leader should ask challenging, probing questions and motivate team members to think deeply and provide honest, unfiltered ideas.

Sometimes encouraging participation is about recognizing when people are hesitant to speak up within a larger group. There are various reasons why a person might not want to share an opinion or idea with a large group but, regardless of the situation, it is the job of the leader to read the room and observe when a participant is uncomfortable. On many occasions I would call breaks or establish smaller, side meetings to tackle issues that were more sensitive or challenging to address, so I could get the unfiltered information I needed without derailing the meeting.

BUILDING A TRUST ENVIRONMENT

Building trust is an essential part of the process of assembling a strong team. Trust affects motivation, engagement, communication, creativity, information flow, and the ability of a team to collaborate effectively. In a complex environment, trust is even more critical because different elements of business and society are so thoroughly intertwined. One person's actions or inactions can ripple out to affect others on the team. In an organization, building trust is a powerful enabler in achieving your mission, while a lack of trust can sabotage your best efforts.

In preparation for writing this book, my Aerospace team members were asked what they considered important to the process of building trust. The examples they gave illustrated fundamental concepts such as empowerment, autonomy, and productive collaboration. As a leader, one way to empower your team is to remember that trust is a two-way street. The leader must set the tone by extending her trust to her collaborators, and her collaborators must learn to trust her in return. Each team member has a unique skill set consisting of specific talents, knowledge, and experience. It's a leader's job to make sure that those skills are being leveraged at the highest level at all times. A great leader is able to get the most out of every team member by emphasizing individual strengths and allowing each person the autonomy to make meaningful decisions and policy recommendations without fear of bias or unjust criticism.

By 2008, Aerospace had a diversified customer base. I knew that my ability to get work done in so many different areas was limited, so I quickly empowered members of my team to interact more directly with customers at the senior level, on my behalf, and with my authority.

My goal was to facilitate a trusting relationship between the customers and the corporate officers, and I had discussions about roles, responsibilities, and expectations with both sides. For example, in 2010 I assigned one of our most experienced senior vice presidents, Dr. David Gorney, the responsibility of managing one of our primary customer portfolios. This customer also happened to be one of Aerospace's neighbors in El Segundo. On the day I gave him the assignment, I walked with him into his office and said, "Look out your window—see that Air Force base? That's your customer now. You now own every aspect of delighting that customer." I then met with the customer and explained why this change in the interface would add value to our business relationship. I assured the customer that they would appreciate the quality time, expertise, and attention the senior vice president would be able to provide. I also let them know that whenever needed, I would be there to provide additional support. If there were hard issues to be worked through, I would still be there, but on a day-to-day basis, the senior vice president was now accountable.

Over time, the trust within our team continued to grow. As my relationships with my direct reports became more established, I re-instituted the executive vice president role and moved Dr. Gorney into the position, where he became a trusted and wise advisor not only to me, but to the rest of the team and to our customers. Now second-in-command, Dr. Gorney enhanced our interactions and improved efficiency across the organization by leveraging the trust and mutual respect he shared with the rest of the team.

The value of a strong "second" to a leader is immeasurable. Typically, that person will wear a lot of hats. He will likely serve as a trusted sounding board for you and the team. Other times, he may take on the role of mentor, coach, or confidant. Ideally, this person keeps his ears and eyes open, routinely scanning the environment to assist with shifting priorities and fast-moving schedules. If you are fortunate enough to have a second with a long tenure at the company, as I did, he can bring a historical perspective to your interactions—illustrating how things were done in the past while pointing out shortcomings that should be remedied in the future. Having this kind of experience to lean on adds a greater sense of stability and continuity to your leadership team. In addition, the support of a strong second enables you to lead on a broader scale and confidently engage with external audiences.

One of the more important qualities to seek in a second is the ability of that person to speak candidly with you. You need to find someone who is not afraid to share his honest opinion, which sometimes means telling you something you might not want to hear. As a leader, and especially as a CEO, you will find it's not always easy to find people who are willing to disagree with you. It's not that people don't disagree with you—believe me, they do—but that they are reluctant to share that dissenting opinion with their boss. It's vital that you can trust your second to give you honest feedback and a clear perspective when others will not. Likewise, it's important that he knows you will hear him without prejudice.

Leaders build trust by listening to what their team has to say, and when practical, implementing or incorporating their ideas into solutions. People pay attention to how leaders treat their individual input and suggestions. Even if an idea is not ultimately used, the fact that it was respectfully heard and considered goes a long way toward ensuring continued participation, not only by the person who offered the idea, but by all those observing the interaction.

Finally, to build trust, it's helpful to remember that your team is made up of human beings. They are not mechanized or infallible. They will make mistakes and create conflict, and that's normal. Each team member has a different set of interests and concerns, and it's very important for a leader to know what those things are, to help their team members learn, grow, and succeed. Leaders can thrive when they establish genuine personal connections with their collaborators. In my experience, the little things go a long way. Thank them publicly when they do well. Learn the names of their family members and pay attention to the milestones they are experiencing in their lives. Congratulate them on birthdays and anniversaries. Support them when a loved one falls ill or passes away. Understand that life is happening all around them and that their work life is only a single part of a much more complex whole.

You are working with, and relying on, these people every single day, so treat them like you care about them. Show respect for their individual concerns and remind them that they are valued. Those moments of appreciation and understanding mean a lot.

DEVELOPING THE TEAM

To have any hope of keeping pace with our rapidly changing society, leaders must actively prioritize and promote learning and development within their organizations. Now more than ever, if we don't learn to learn, we run the risk of being quickly left behind. As a leader, you never stop developing. Leadership is a constant process of reinvention and refinement. The world around you never stays the same, so neither can you.

The process of consistently building the capacity for your organization to evolve and adjust is what Harvard professor Ronald Heifetz refers to as "adaptive leadership." In his book *The Practice of Adaptive Leadership*, Heifetz writes that "adaptive challenges can only be addressed through changes in people's priorities, beliefs, habits, and loyalties. Making progress requires going beyond any authoritative expertise to mobilize discovery, shedding certain entrenched ways, tolerating losses, and generating new capacity to thrive anew."

Essentially, Heifitz is saying that we need to jump right in and learn our way through situations we've never faced before. For example, the expansion of the Internet has completely transformed our society, creating many new opportunities but also a host of challenges and security concerns that must be addressed. As a global society we are constantly evolving, and in order to survive these evolutions, we need to keep changing and adapting.

When I began as CEO, I was unaware of the magnitude of change I would experience. At Aerospace, I had steadily moved up the ranks, serving in two senior vice president roles. As prepared as I thought I was, the CEO role still included non-stop development from my first day on the job to the very last. Every day I learned something new and became more adept at handling adaptive challenges.

An early development challenge for me was learning to effectively manage a much larger portfolio of customers and external stakeholders than ever before. This didn't just require me to build more technical knowledge. This was an adaptive challenge, requiring a shift in priorities and an expansion of my loyalty to many new employees and customers. To make it all possible, I knew I needed to rely heavily on my team for expertise, guidance, and practical support. I made sure that my team was comfortable working directly with our customers and vice versa. I made sure to connect the team with the board of trustees in a way that encouraged more frequent communication and collaboration. We also worked on building mutual trust by helping the board members learn more about the Aerospace officers' various roles, through casual conversations and purposeful interactions.

My development strategy was to strengthen the sense of "we" within my team, by creating a common vision, sense of ownership, and operating principle that guided all of our interactions. This represented both a significant increase in collective capability and a shift in the way that we worked together. It was an improvement, but it also changed how we conducted business. Any time you seek to think and operate differently, you will face challenges that are an inherent part of the learning and development process. To implement change, it is critical to always make space for assessment, self-evaluation, and course-correction. You need to take time to acknowledge "lessons learned" at the completion of a project and make sure those lessons guide your future actions. We all want to know what we did well, but it's also important to be clear about what we should do differently next time.

If you've been a leader for any length of time, it's likely you have already identified effective ways to develop your staff. Here are a couple of techniques I've found work well:

First, I like to expose my thought process to my staff through conversation, rather than through directives or orders. "Teachable moments" can occur during any interaction, in any setting or context. They allow my team to know not only what I'm thinking, but how and why. By taking the extra time to share your thoughts, you provide the opportunity for team members to correct an error in their thinking or add additional facts to the conversation. It can also increase the likelihood that in a similar situation in the future, your team will make decisions consistent with your own expectations and preferences.

Second, I look for ways to support my team members in the examination of their own thoughts and ideas. The tendency for most people is to assure their boss that they have everything under control. This tendency is not helpful to you as a leader, especially when it's not true. It takes courage for any of us to be transparent in our thinking. By serving as a sounding board, a leader can stretch her team's thinking and allow them to open up. You can help them consider new perspectives and approaches that may be outside the scope of their current thinking. You can help them grasp the bigger picture by sharing what you know about other team members' opinions. You can also ask them to think about what they would have done differently if they could do the just-completed project over again.

In a work setting, most people tend to focus on one specific part of the job—an individual piece of the puzzle. As a leader, you're exposed to more of those pieces, and when you share your unique insights, the team is able to make more robust decisions. Anyone on the team can boost the group's effectiveness by pointing things out, raising questions, or offering to assist with contacts or networks. Sometimes all that is needed is to let your team think out loud and explore new ideas without fear of reproach. You want to empower creativity and new ideas. You want your team to be fearless and transparent in their pursuit of the best solutions to your organization's toughest problems.

84

Stretch Assignments

Understanding the thoughts and concerns of your team members is very important, but it's equally important to challenge and push them to take on new assignments that can broaden and deepen their individual skill sets. One of my favorite ways to do this was to give aspiring leaders throughout my organization "stretch assignments." These assignments allowed team members to explore new areas of expertise and to take on different responsibilities than they were typically accustomed to. For example, in 2015 we hosted a science, technology, engineering, and math (STEM) event, which brought together local industry, academia, and community leaders to discuss ways to better promote and improve STEM literacy in America. I selected two senior-level technical managers to chair the event as a stretch assignment for both of them.

These managers were accustomed to working exclusively on the technical end of our business, assisting in the support of on-orbit space assets and the launch of rockets. A social event designed to spur community engagement was not a typical business responsibility for them, but I saw tremendous value in exposing them to a new network of people with diverse interests and concerns. Supporting the STEM event, these managers were forced to learn new things, ask for advice from experts, execute a large event plan, and, most importantly, to lead without being subject-matter experts.

This assignment placed these two technical experts in a vulnerable situation, where they had to rely on others for insight and guidance. They had to learn to think like leaders and see the big picture while relying on the expertise of others. It was a great challenge for them and, as a result, they gained valuable new leadership skills that could be applied to all of their future work assignments—technical or not.

I also found that stretch assignments were particularly valuable at the officer level. Like many companies, Aerospace frequently received invitations for leaders to participate in conferences and panel discussions. I often volunteered Aerospace officers to participate in these conferences as stretch assignments. This would temporarily burden the individuals with additional responsibilities, but the benefits of their participation always outweighed the temporary inconvenience of an increased workload. Conference participation and other public engagement opportunities can broaden the understanding of what it means to represent a company, providing a chance to expand networks, gain new insights, and make a significant contribution to the industry as a whole.

I often delivered speeches and attended events to build relationships within the community and strengthen our corporate presence on a larger scale. Unfortunately, my unpredictable schedule and frequent last-minute business commitments made it impossible to accept every invitation I received. In those instances, I delegated other corporate officers to attend, using those events as opportunities to develop my team and expose others to the depth of our leadership bench.

Often, when I sent an officer to appear at an event in my place, my suggestion was met with some trepidation. Either the officers were unfamiliar with the audience, or they had never spoken to such a large group before, or they were concerned about their ability to perform and represent Aerospace at a high level. In these instances, I always made sure to empower them, explain that we shared a collective vision for our company, and that their intuitive understanding of the corporation's position on significant issues would allow them to speak with confidence. Moreover, I would make it very clear that I was committed to their success and was prepared to invest in any media training and coaching they might need.

SUCCESSION PLANNING

It seems fitting to close this chapter with a few comments about succession planning. An important part of leadership is leaving your organization in a position to thrive after your departure. When you walk out the door, the wheels of business must continue to turn. A new leader will take over and make changes, but it's your responsibility to provide that leader with a team, an infrastructure, and a culture that is primed for future success.

At every quarterly board meeting at Aerospace, we made sure to review the succession plan for all of our officer positions. In addition, twice a year, my direct reports and I evaluated and planned for the succession of *their* direct reports.

Succession planning and strategic talent development go hand in hand; both are critical parts of a senior leader's role. The challenge with succession is to look broadly at the company's total leadership pool and to consider longer-term talent and skill needs. In addition to preparing to fill anticipated vacancies and developing current leaders, you also need to anticipate the skills that will be needed for *their* successors. You will face the challenge of balancing the development needs of corporate-level talent pools with the immediate needs of the local hiring manager.

As a senior leader, you have a responsibility to build your company's potential leadership talent pool. You have to recognize who is available within your company, what their skill sets are, and what the ramifications are of moving employees into new roles, and, potentially, new areas of the company. We recognized that talent development should be a critical part of our everyday operations, so some portion of every staff meeting was focused on making promotion decisions, identifying talent for stretch assignments, and/or reviewing progress on talent development efforts. We wanted to make sure that we were preparing the right people for the right roles at all times.

There are only so many advancement opportunities within a company, and the higher up you go, the fewer the options are. Part of job placement is timing, but part of it is also proactively finding or creating opportunities that allow a high-potential leader to develop the skills and experiences she needs to take on more significant roles.

Ultimately, when promotion decisions are made, they are based on a blend of candidate readiness and opportunity. Leadership development also requires that you make space for it. It has to be practiced and intentional every step of the way, to ensure that when a position opens up, the right leader is prepared to step into the role.

Not all positions at the same level are equal. Some key positions are more representative of a skill set that will be needed at a higher level than others. The significance of a position can be impacted by the level of customer interface required, the geographic dispersion of the organization, and/or the size of the organization. When making a selection for a key position, it's important to look carefully at the leadership characteristics and emotional intelligence that the job requires in addition to its basic technical requirements. Sometimes a difficult choice between two equally qualified candidates might come down to which person has shown the higher potential for leadership development.

The logic behind this approach is universally applicable—weigh the pros and cons, and choose the candidate with the most upside. Every hiring decision has risk because there are no guarantees. It's important to remember that every time you promote someone, you have made a decision to invest in the success of that individual. To ensure that success, the entire leadership team must work together to provide the mentoring and resources needed to give the selected candidate every opportunity to flourish. It's also important for employees in new positions to have enough humility to be willing to learn and recognize that they need the team's support to help them be successful.

One of the unique things about being CEO is that you are responsible for understanding the entire scope of your business. If you have a candidate who shows the capacity and learning agility to do more than her current role allows, you need to help the candidate see the company from a more expansive perspective. It's rare to find a candidate who understands the potential she has for her own career. Having discussions with those individuals about their strengths, their weaknesses, and their untapped talents is very instructive. It is also important that they be given the opportunity to do the hard work that is required to develop themselves. You need to let these individuals know that there are opportunities for them to grow and excel, but that those opportunities depend largely upon their own willingness to put in the requisite effort.

Throughout this chapter, I've suggested several ways to work with your team, both in the short term and the long term. One thing is certain—change is inevitable, and that means you and your team will need to continually assess your alignment with a dynamic set of goals and opportunities. You will all need to continue to adapt in collective response to the changing environment. You will all need to sustain and protect that valuable sense of "we" by maintaining a common vision and operating principle. Learning is crucial, as is developing ways to see more of what's happening through networks and asking the right questions.

The work environments you create are also extremely important. To perform at a high level, you need to trust and respect your team members and they need to do the same for you. Everyone needs to be on the same page, and when it clicks, the team becomes a force multiplier, resulting in exquisite accomplishments for your company. Teamwork, when it is working well, is a work of art.

CHAPTER SEVEN

Lessons Learned

BRINGING IT ALL TOGETHER.

If you are fortunate enough to find yourself in a leadership role in any area of life or business, you will have the opportunity to influence the way that people think and act in order to achieve results. Being a leader is a tremendous responsibility, but the responsibility that comes with leadership is a gift because it allows you to improve the lives of others and accomplish great things on a large scale. By embracing your leadership role, you empower your colleagues and collaborators, and in so doing, empower yourself.

When I started my career in engineering, I had no idea that I would eventually find myself in a significant leadership role. There were so many learning experiences along the way, and every time I thought I had reached a plateau, I would find an unexpected challenge to overcome and a unique skill to learn. I constantly had to evolve and refine my approach to leadership. There was always something new to learn and someone new to learn from.

In 2015, I was appointed to the President's Council of Advisors on Science and Technology (PCAST). PCAST is an advisory group consisting of leading scientists and engineers who directly advise the President and the Executive Office of the President. On the day of my first meeting, I walked into the Roosevelt Room in the White House and took my seat at the table with the other PCAST members and members of the White House leadership. After a few moments of excitement and anticipation, President Obama walked into the room. He briefly greeted every individual at the table and immediately engaged with the group on a range of varied subjects.

I listened carefully and participated when I could. I was amazed at the President's depth of knowledge and the way he enthusiastically led our discussion about science and technology with relative ease and efficiency. He appeared focused and engaged at all times, and he made everyone in the room feel as though this meeting was the most important thing that he had to do that day—a remarkable feat given the range, magnitude, and complexity of the issues that the President works on every day. He brought the best out of the group and demanded actionable options at all times.

On my flight home after the meeting, I took the opportunity to reflect on my experience. I had just spent an hour with the President of the United States, the leader of the free world, and had witnessed something remarkable—he had a thirst for knowledge. At his level, every issue is complex and every decision has strategic implications, yet he was still making sure to exercise his information network so as to increase his personal knowledge and set the direction for this team. He was still pushing himself and those around him to take on challenges and question time-tested procedures and concepts. His leadership style was inspirational. In that moment of reflection, I realized that I too would never, and should never, cease to push myself to hone my abilities as a leader. Leadership development truly is a never-ending process, and it's a process that rewards effort—the more you put into it, the more you get out of it.

Throughout this book, I've shared my perspective on the elements that are critical to being a successful strategic leader in a complex world. Everything I've learned about leadership has resulted from meaningful interactions with my fellow employees, colleagues, and mentors throughout my life. Sometimes we gain new skills and discover new approaches, but forget them because we don't put them into practice. This book is designed to help leaders address that problem by distilling a number of important lessons and strategies into tactics that can be implemented in small or large doses. It is by no means comprehensive, but my hope is that it can serve as a guide for leaders and aspiring leaders as they evolve and grow in their professions.

In closing, I'd like to emphasize that leadership is not a birthright; it is a skill. Leaders can come from anywhere and in any form. There was nothing in my background to suggest that I would be able to achieve what I have in my career. There was no template for me to follow; in fact, there was a societal assumption that an African American woman from the inner city in the 1960s could not be a leader. I proved that assumption wrong. I relied on the encouragement and mentorship of others, and I relied on my work ethic and values to develop my skills, and eventually become the leader I am today. I have learned, first-hand, that great leadership is not inherited—it is earned and it is learned, one lesson at a time.

Appendix

Leadership Maxims

Below, I present some of my observations and lessons on leadership in short form that make a point and remind me that leadership isn't easy. They are not unique to the aerospace industry, but rather provide food for thought for aspiring leaders at every level.

- Leadership doesn't just happen. You have to make it a priority and commit to making space for it to happen.

- A leader brings solutions, not problems, to the table; if you are not part of the solution, you are part of the problem.

- Instead of becoming paralyzed by watching the door that is closing, keep watching for the one that is about to open.

- Knowledge is not leadership and intelligence is not experience.

- If you think getting promoted is the answer to your problem, think again.

- Anticipate change and harness it. We live in an age of relentless change. Remember: what you thought you knew yesterday may indeed be irrelevant today.

- Be proactive with your team on their personal and their career development.

- It's great to know your strengths, but it's critical to work on recognizing and overcoming your weaknesses.

- If you are not having a crisis today, just wait a minute– you will.

- When you find yourself standing in quicksand, don't respond by flailing around. Have a backup plan that will get you back to terra firma in a safe, stable, and sustainable way.

- You are not the boss. There are important stakeholders in every circumstance; at best, you are the conductor trying to keep everyone on the same piece of music, on the beat, and in the right key.

- Ask good questions and then listen carefully to the answers for what is said and what is not said.

- Be truthful and transparent to the greatest extent possible.

- Mentoring is an essential part of your role; look for teachable moments and provide feedback often.

- Connect with the members of your team on a personal level. Show respect for them and they will show respect for you.

- Be willing to take measured risks in the interest of innovation and continuous improvement.

- Not making a decision IS a decision.

- Admiring the problem is not the same as working to resolve it.

- Leave things better than you found them.

BIBLIOGRAPHY

HBR's 10 Must Reads on Leadership, Boston, MA, Harvard Business Review Press, 2011.

The CEO Report: Embracing the Paradoxes of Leadership and the Power of Doubt, Heidrick & Struggles and Saïd Business School at the University of Oxford, 2015.

Armour, Vernice. *Zero to Breakthrough: The 7-Step, Battle-Tested Method for Accomplishing Goals That Matter*, New York, NY, Gotham, 2011.

Austin, Jan. *What No One Ever Tells You About Leading for Results: Best Practices from 101 Real-World Leaders*, Chicago, IL, Kaplan Publishing, 2006.

Austin, Wanda Murry. *A Methodology for System Dynamics Modeling Using Natural Language*, Doctor of Philosophy Dissertation, Industrial and Systems Engineering, University of Southern California, 1988.

Bossidy, Larry; Ram Charan; and Charles Burck. *Execution: The Discipline of Getting Things Done*, New York, NY, Crown Business, 2002.

Burnison, Gary. *The Twelve Absolutes of Leadership*, New York, NY, McGraw-Hill, 2012.

Charan, Ram; Stephen J. Drotter; and James L. Noel. *Leadership Pipeline: How to Build the Leadership-Powered Company*, San Francisco, CA, Jossey-Bass Inc., 2001.

Clemons, Hank. *Leadership and the Influence of Diversity: What Every Leader Ought to Know*, Tampa, FL, HLC Group Inc., 2003.

Collins, James C. *Good to Great: Why Some Companies Make the Leap—and Others Don't*, New York, NY, Harper Business, 2001.

Conant, Douglas R., and Mette Norgaard. *Touchpoints: Creating Powerful Leadership Connections in the Smallest of Moments*, Douglas R. Conant, Mette Norgaard, Warren Bennis signature series, 1st ed., San Francisco, CA, Jossey-Bass Inc., 2011.

Cottrell, David. *12 Choices ... That Lead to Your Success*, Dallas, TX, CornerStone Leadership Institute, 2005.

Covey, Stephen R. *Principle-Centered Leadership*, New York, NY, Summit Books, 1991.

Daly, Peter H.; Michael Watkins; and Cate Reavis. *The First 90 Days in Government: Critical Success Strategies for New Public Managers at All Levels*, Boston, MA, Harvard Business Press, 2006.

Day, George S., and Paul J. H. Schoemaker. *Are You a 'Vigilant Leader'?* Cambridge, MA, MIT Sloan Management Review, 2008.

Donlon, J.P. *How CEOs Can Get It Right,* Accessed 02/22/2016, http://chiefexecutive.net/how-ceos-can-get-it-right/, 2013.

Gladwell, Malcolm. *Outliers: The Story of Success*, New York, NY, Little, Brown and Co., 2008.

Goleman, Daniel. *What Makes a Leader?* Boston, MA, Harvard Business Press, 2008.

Heifetz, Ronald A.; Alexander Grashow; and Martin Linsky. *The Practice of Adaptive Leadership: Tools and Tactics for Changing Your Organization and the World*, Boston, MA, Harvard Business Press, 2009.

Kahneman, Daniel. *Thinking, Fast and Slow*, Farrar, Straus and Giroux, New York, NY, 2011.

Keller, Gary, and Jay Papasan. *The One Thing: The Surprisingly Simple Truth Behind Extraordinary Results*, Austin, TX, Bard Press, 2013.

Lombardo, Michael M., and Robert W. Eichinger. *FYI for Your Improvement: A Guide for Development and Coaching for Learners, Managers, Mentors, and Feedback Givers*, South Minneapolis, MN, Korn Ferry, 2009.

McChrystal, Stanley A.; Tantum Collins; David Silverman; and Chris Fussell. *Team of Teams: New Rules of Engagement for a Complex World*, Penguin Publishing Group, New York, NY, 2015.

Muson, Howard. *How CEOs Turn Themselves (and Others) into True Leaders*, Executive Action Report, New York, NY, The Conference Board, Inc., 2008.

Presley, Stephen P. *How Leaders Engage in Complexity Leadership: Do Action-Logics Make a Difference?* Doctor of Philosophy Dissertation, The School of Human and Organization Development, Fielding Graduate University, 2014.

Pritchett, Price. *The Ethics of Excellence*, Dallas, TX, Pritchett Publishing Co., 1991.

Sample, Steven B. *The Contrarian's Guide to Leadership*, San Francisco, CA, Jossey-Bass Inc., 2002.

Salovey, P. and J.D. Mayer. *Emotional Intelligence, Imagination, Cognition and Personality,* Amityville, NY, Baywood Publishing Company, Inc., 1990.

Sandberg, Sheryl. *Lean In: Women, Work, and the Will to Lead*, New York, NY, Knopf, 2013.

Senge, Peter M. *The Fifth Discipline: The Art and Practice of the Learning Organization*, New York, NY, Doubleday/Currency, 1990.

Stamoulis, Dean T. and Erika Mannion. *Making It to the Top: Nine Attributes That Differentiate CEOs*, In Touch with the Board, Russell Reynolds Associates, New York, NY, 2012.

Sterman, John D. *Learning from Evidence in a Complex World*, American Journal of Public Health, 2006.

Tichy, Noel M. and Warren G. Bennis. *Making Judgment Calls: The Ultimate Act of Leadership*, Boston, MA, Harvard Business Review, 2007.

Weisgerber, Marcus, *Hagel: Budget Uncertainty Is the Biggest Challenge Facing the Military*, defenseone.com Accessed 02/23/2016, http://www.defenseone.com/management/2015/01/hagel-budget-uncertainty-biggest-challenge-facing-military/103065/. 2015.

White, J. D. and L. G. Tilney. *Introduction to the Test-Like-You-Fly Process, Parts 1 & 2*, Briefing presented at the 26th Aerospace Testing Seminar, The Aerospace Corporation, El Segundo, CA, March 2011.

Young, Zoe. *Paul Ignatius Lectures Students on Leadership*, Daily Trojan, September 17, 2013, http://dailytrojan.com/2013/09/17/paul-ignatius-lectures-students-on-leadership/.

ABOUT THE AUTHOR

Dr. Wanda M. Austin served as the president and CEO of The Aerospace Corporation from 2008 to 2016. She was a member of the technical staff at Aerospace for nearly 29 years prior to leading the company. She is an award-winning engineer and mathematician who has served on many technical boards, including the Defense Science Board, the Air Force Scientific Advisory Board, and the President's Council of Advisors on Science and Technology. In addition, she has been a trustee of the University of Southern California, a member of the board of the National Geographic Society and The Space Foundation, an honorary fellow of the American Institute of Aeronautics and Astronautics, a councilor of the National Academy of Engineering, and a member of the American Academy of Arts and Sciences. Her reflections on leadership are drawn from both her professional career and her experiences as a daughter, wife, mother, and grandmother.

Made in the USA
San Bernardino, CA
12 July 2017